Integration of Legume Crops with Cereal Crops Under Changing Climate

Samiha Ouda · Abd El-Hafeez Zohry

Integration of Legume Crops with Cereal Crops Under Changing Climate

Sustainably Increasing Food Production

Springer

Samiha Ouda
Water Requirements and Field Irrigation Research Department
Agricultural Research Center, Soils, Water and Environment Research Institute
Giza, Egypt

Abd El-Hafeez Zohry
Crops Intensification Research Department
Field Crops Research Institute
Giza, Egypt

ISBN 978-3-031-68101-1 ISBN 978-3-031-68102-8 (eBook)
https://doi.org/10.1007/978-3-031-68102-8

This Springer imprint is published by the registered company Springer Nature Switzerland AG
The registered company address is: Gewerbestrasse 11, 6330 Cham, Switzerland

Dedicated

To those who work in the shadow for the greater good

Preface

"Peace and Prosperity for People and the Planet"

This is the aim of the 2030 Agenda for Sustainable Development, adopted by all United Nations members in 2015. It created 17 world Sustainable Development Goals (SDGs) to attain this aim. The SDGs highlight the connections between the environmental, social, and economic aspects of sustainable development. The title of the Goal 2 has stated "end hunger, achieve food security and improved nutrition and promote sustainable agriculture." Because sustainability is at the center of the SDGs, the sustainable use of soil and water resources to produce food is one of the most important milestones to achieve. Furthermore, combating the detrimental effect of climate change on food production necessitates the implementation of smart practices that can mitigate its harmful effects.

Cereal crops are one of the important food classes that provide mankind with energy and protein requirements as well as contribute in attaining food security. However, the continuous cultivation of cereals causes soil deterioration. To overcome this setback, the integration of legume crops within cereals-based cropping system can play a vital role in reducing the deterioration in the soil through the symbiotic fixation of nitrogen in the roots by rhizobium, which increases soil nitrogen availability to the following crop, as well as rise soil organic content. Furthermore, legumes mobilize phosphorus in the soil during its growth, which increases its uptake by the following cereal crop. It has also been reported that cultivation of legume crops enhances soil water holding capacity and water retention. Thus, including legume crops within the cereals-based cropping system can attain the sustainable use of soil and water resources, as well as increase the productivity of cereal crops.

In this book, we will thoroughly tackle the benefits of inclusion of legume crops within cereal-based cropping system, namely wheat, maize, and rice (paddy and upland) under the changing climate (current and future). We will review the innovations and interventions that could sustainably intensify the production of cereals to reduce hunger and poverty.

The book consists of an introduction and five chapters as follows:

Chapter 1 will discuss two approaches to face the shortage in wheat production and fulfill the existed demand worldwide. Breading for more resilient wheat cultivars to face biotic (viruses, fungi, and insects) and abiotic (heat, drought, and salinity) stresses, as well as sustainably increase in wheat production are the two suggested approaches to globally reduce wheat production-consumption gap. Globally, breeding programs for resilient wheat cultivars have been implemented and identified genetic resistance against biotic stress, as well as genetic mechanism caused by abiotic stresses. Additionally, breeding for improved root system of wheat was done to improve water and nutrients uptake. With respect to the second approach, namely sustainable increase wheat production, increasing nutrients (nitrogen, phosphorus, and potassium) use efficiency will be reviewed, as well as reducing the reliance on chemical fertilizer by application of organic fertilizer. The effect of rotating wheat with legume crops, and implementing wheat intercropping system with legume crops will also reviewed. Moreover, increasing water use efficiency and water productivity can also conducted to sustainably increase wheat production through improving cultivation method and using more efficient irrigation system.

Chapter 2 will review the Egyptian breeding efforts in producing resilient wheat cultivars to face biotic and abiotic stress. Several improved management practices to increase wheat production will be also reviewed, namely reducing the reliance on chemical fertilizer in wheat production by application of organic fertilizer, legume crops preceding wheat, and wheat intercropping systems with legume crops. Moreover, the advantage of intercropping fahl clover with wheat will be presented in a field study, where fahl clover was intercropped with wheat, in three planting densities and under three cultivation methods.

Chapter 3 analyzed the relationship between some weather elements (solar radiation, maximum and minimum temperatures, and relative humidity) and maize productivity in the five agro-climatic zones of Egypt from 2009 to 2021 using multiple linear regression analysis. Furthermore, the projected effect of climate change on maize productivity was assessed in the five agro-climatic zones of Egypt using CMIP6 mean projection of two future scenarios, namely SSP1-2.6 and SSP5-8.5 in two time-intervals (2060–2079 and 2080–2099). Additionally, a brief review will be presented on the effect of intercropping legume crops with maize on relieving the consequences of climate change on maize productivity.

Chapter 4 is composed of two sections: In the first section, we will test the capability of SALTMED model to simulate multi-year crop sequences of maize and wheat, as well as simulation of the inclusion of legume crops, namely fahl clover and cowpea within the cropping system of maize and wheat in Egypt. In the second section of this chapter, we will evaluate the role of legumes in enhancing the resiliency of maize and wheat to climate change under the studied crop sequences using SALTMED model and CMIP6 mean projection of two future scenarios, namely SSP1-2.6 and SSP5-8.5 in two time-intervals, namely 2060–2079 and 2080–2099.

Chapter 5 in this chapter we will use a modeling approach to assess the impact of future climate change on water requirements and water productivity of three cultivars of paddy rice planted in a two-year field experiment in Egypt under future climate scenarios developed by CMIP6, namely SSP1-2.6 and SSP5-8.5 in two time-intervals, namely 2060–2079 and 2080–2099. We will also employ a simulation approach using SALTMED model to assess the impact of climate change under the above scenarios on the yield of upland rice grown under deficit irrigation. Furthermore, we will review the attempts to intercrop legume crops, namely peanut, mung bean, and soybean with upland rice to increase its productivity in a sustainable manner. These intercropping systems proved to be effective in increasing the yield of upland rice, and it could be implemented in the future to reduce the harmful effect of climate change on rice production.

We hope that this book will assist researchers in other developing countries in their sensible missions to improve their situation of food security to attain better life and prosperity for their citizens.

Giza, Egypt
2024

Samiha Ouda

Introduction

The world population steadily increases with high rate in the past twelve years from 7126 billion inhibitors in 2012 (FAOSTAT, https://www.fao.org/faostat/en/#data) to 8095 billion inhibitors in 2024 (Worldometer, https://www.worldometers.info/world-population), which amount to 14% increase. In the meantime, the number of severely food insecure people was 604.5 million in 2014, and it increased by 53% in 2020 to reach 927.6 million (FAOSTAT). These numbers raise large concerns about the future of food production to feed this continually growing population. Lately, many developing countries rely on importing large quantities of crops, such as wheat, maize, rice, and barley to meet their food and feed needs. The negative impact of climate change and its consequences, namely low crops' productivity and water scarcity causing great disruptions in food production systems.

As changing in the climate of the earth has become more pronounced, more resilient environmental systems to these changes should be developed. The Fourth Assessment Report of the Intergovernmental Panel on Climate Change (IPCC, 2013) stated that "the magnitude of stress on water resources is expected to increase as a consequence of climate change, in addition to future population growth and urbanization as a consequence of land-use change." Although climate change is a global phenomenon, its impacts are more displayed in the developing countries, as a result of their higher vulnerabilities and lesser ability to adapt because they are agriculture-based economies affected by direct exposure to nature (Mendelsohn, 2014). Climate change phenomena have many syndromes, i.e., high temperature and variable rainfall, which manifest as extreme weather events. According to the scenarios described in the IPCC Special Report on Emissions Scenarios, changes in precipitation and temperature may lead to changes in runoff and water availability (Cisneros et al., 2014). Furthermore, climate change negatively affects the agricultural sector through its impact on crops' productivity, its stability, and other components of the food system, including storage, access, and utilization (Wheeler et al., 2013). It was estimated that up to one-fifth of the global population could suffer severe shortages of freshwater (Schiermeier, 2014) or its quality in the future (Girones et al., 2010). Furthermore, FAO (2012) predicted an increase in food demand by 50% and an increase in water use in irrigation by 30% in 2030.

To improve the situation of global food security, the Food and Agriculture Organization of the United Nations developed the concept of "climate-smart agriculture" to achieve one of the important goals of the 17 SDGs, namely Goal 2, which pursues to end hunger and ensure access of poor people and those in vulnerable situations, to safe nutritious and sufficient food all year round by the year 2030. Target 4 of this goal seeks to ensure sustainable food production systems and implement resilient agricultural practices to climate variability that could increase crops' production and reduce food insecurity. The most important food classes are cereal and legume crops, which provide humans with nutritious sources of food more than any other food categories worldwide.

Global Cereal Crops

Cereal grains provide humankind with more nourishment, namely energy and protein requirements than any other food classes. They are a rich source of vitamins, minerals, carbohydrates, fats, oils, protein, fiber, minerals, β-glucan, lignans, phytosterol, phenolics, and among others. These compounds provide health benefits on consumption, such as antioxidants, antidiabetics, antihypertension, antimicrobial, and anticancer with no side effects (Oluwajuyitan et al., 2021). All cereal grains have high energy values, mainly from the starch fraction and protein portions. The carbohydrates mainly starch contents in the cereal grains comprising 65–75% of their total weight, proteins comprising 6–12%, and fat comprising 1–5% along with traces of minerals and vitamins (Sarwar et al., 2013).

Although there are many cereal crops used for food, only three of them are considered the most important human food sources, namely wheat, maize, and rice (Table 1). Wheat is highly consumed in China, European Union, and India, and maize is highly consumed in the USA and China. Whereas, rice is highly consumed in China and India. Globally, wheat occupies 30% and maize occupies 28%, whereas rice occupies 23% of the total cultivated area of cereal crops. The highest global production of a cereal crop is for maize, namely 1162.4 billion tons, followed by wheat and rice, namely 760.9 and 756.7 million tons, respectively (FAOSTAT).

On the other hand, the global cultivated area of the other cereal crops presented in Table 1 is considered low, compared to these three main crops, namely barley, sorghum, and millet. These three crops are less popular, where its cultivated area is lower. Furthermore, the lowest cultivated areas were found for oats and triticale. This could be due to the unsuitability of these crops for cultivation in some geographic zones, which requires cool weather. Oats are particularly well-suited in areas with cool, wet springs (Savova et al., 2002). Whereas, triticale is cultivated on a global scale with the main production areas in Central and Eastern Europe, where the largest producers are Poland, Germany, France, and Russia (FAO, 2015).

Global production of wheat and maize is the most affected cereal crops by the Russian aggression against Ukraine, as a result of no grain exportation. Whereas,

Table 1 Global average productivity, cultivated area, and total production of the main cereal crops in 2020

Crop	Productivity (ton/ha)	Cultivated area (000,000 ha)	Total production (000,000 ton)
Wheat	3.47	219	760.9
Maize	5.75	202	1162.4
Rice	4.61	164	756.7
Barley	3.04	52	157.0
Sorghum	1.46	40	58.7
Millet	0.95	32	30.5
Oats	2.58	10	25.2
Triticale	4.03	4	15.4

Source FAOSTAT

the productivity of the other cereal crops will be affected through unavailability of fertilizer and fuel.

Global Legume Crops

On the other hand, legume crops are considered the second most important food source after cereals. Legume crops are varying nutritionally food, providing 20–45% of proteins that contain essential amino acids and vitamins. It also provides dietary fiber and is low in fat, with the exception of soybeans and groundnut, which contains 47 and 45% fat. Moreover, previous research has reported that legume crops have antioxidant properties, which play a role in the prevention of some cancers, heart diseases, osteoporosis, and other degenerative diseases (Maphosa & Jideani 2017).

Additionally, legume crops can be successfully grown in poor soils and adverse weather conditions, in addition to being highly disease and pest-resistant. It can be also used as cover crops, which reduce soil erosion (Karmakar et al., 2016). Its symbiotic relationship with the nitrogen-fixing rhizopus resident in their root nodules makes them excellent crops to improve soil organic matter (Cagasan & Amarado 2023).

There are six main legume crops prevailing worldwide (Table 2). Of them, soybean occupies the highest cultivated areas and total production, namely 126.95 million hectares, which amounted to 59% of the total cultivated areas with legumes. The soybean-cultivated areas produced 353.5 million tons. Beans occupy the second highest cultivated areas followed by groundnuts, which contributed with 16 and 15% of the total cultivated areas with legumes. Soybean is highly consumed in China, and the USA, and groundnut is highly consumed in China and India. Whereas, beans are highly consumed in India and Brazil (FAOSTAT). Other legume crops presented

Table 2 Global average productivity, cultivated area, and total production of the main legume crops in 2020

Crop	Productivity (ton/ha)	Cultivated area (000,000 ha)	Total production (000,000 ton)
Soybeans	2.78	126.95	353.5
Groundnuts	1.70	31.57	53.6
Beans	0.79	34.80	27.5
Chickpeas	1.02	14.84	15.1
Lentils	1.30	5.01	6.5
Lupines	1.18	0.89	1.0

Source FAOSTAT

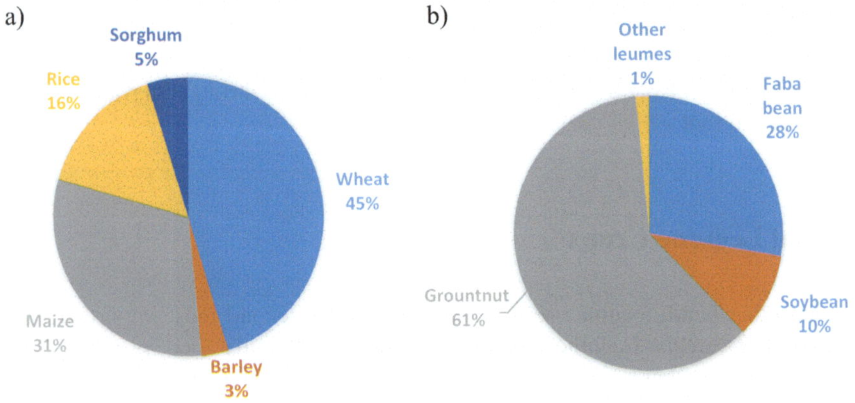

Fig. 1 Percentage of the cultivated area of popular cereal crops (a) and legume crops (b) in Egypt in 2019/20 growing seasons

in Table 2, namely chickpeas, lentils, and lupines occupy lower cultivated areas, compared to the above-mentioned legume crops.

The Main Cereal and Legume Crops in Egypt

Cereal and legume crops are very important food groups in Egypt. The most cultivated cereal crops classified by its cultivated areas in 2019/20 growing seasons (Fig. 1a) were wheat, which occupies the highest area, followed by maize and rice. Both sorghum and barley occupy smaller cultivated areas, namely 5 and 3%, respectively.

The Egyptian cultivars of wheat and maize are suited to be cultivated in all the Egyptian governorates from north to south. Rice is cultivated in north Egypt's governorates and in the salt-affected soils of governorates. Barley cultivars are also suited to be cultivated in all Egyptian governorates; however, its cultivated areas are low. Whereas, sorghum is cultivated in south Egypt (CAPMAS, 2020). Egypt is characterized by having low self-sufficiency ratios in several commodities. Wheat self-sufficiency ratio is 43%, maize self-sufficiency ratio is 47%, and barely self-sufficiency ratio is 81%. Whereas, Egypt is self-sufficient in rice and sorghum.

With respect to legume crops (Fig. 1b), groundnut represented the highest cultivated area, namely 61% of the total cultivated area of legumes. Faba bean and soybean represented low value of the cultivated area with legumes, namely 28 and 10%, respectively. Other popular legume crops, namely chickpeas, lentils, and lupines, occupy only 1% of the total cultivated area of legume crops (CAPMAS, 2020). Lower self-sufficiency ratio existed in the popular legume crops, namely 20% for faba bean, 4% for soybean, and 36% for other legumes (lentil, chickpea, and lupine). Whereas, Egypt is self-sufficient in groundnut.

Increasing the production of cereal and legume crops worldwide can be achieved through increasing average yield per unit area or expanding the area devoted to cereals and legumes into more marginal lands. Moreover, breeding for more resilient cultivars, which can release its potential yield could play an important role in increasing its total production under adverse growth conditions.

Furthermore, inclusion of legume crops, such as soybean, faba beans, and cowpea in cereal-based cropping systems is a viable strategy to reduce inorganic fertilizer use (Danga et al., 2009). Several studies reported savings between 100 and 200 kg/ha in the amount of nitrogen fertilizers applied to cereal crops following faba bean (Preissel et al., 2015; Nuruzzaman et al., 2005). Furthermore, increase in the production of cereal crops was reported when legume crops were included in cereal-based cropping systems (Singh et al., 2019). Whereas, Tang et al. (2014) reported an increase in soil nitrogen, phosphorus, and potassium when legume crops were included in cereal-based cropping systems.

In this book, we thoroughly tackled the benefits of integration of legume crops within cereal-based cropping systems, namely wheat, maize, and rice (paddy and upland) under the changing climate (current and future). We also reviewed the innovations and interventions that could sustainably intensify the production of cereals to reduce hunger and poverty. We used both modeling and simulation approaches to assess the impact of climate change using CMIP6 mean projection of two future scenarios, namely SSP1-2.6 and SSP5-8.5 in two time-intervals (2060–2079 and 2080–2099) on the yield and water requirements of wheat, maize, and rice (paddy and upland).

References

Cagasan, U., & Amarado, J. (2023). Influence of intercropping on the growth and yield performance of upland rice (Oryza sativa L. var. zambales) and peanut (Arachis hypogaea L.). *Research in Agricultural Sciences, 54*(1), 2–8.

CAPMAS. (2020). *Annual Statistical Year Book 2020.* Central Agency of Public Mobilization and Statistics.

Cisneros, J. B. E., Oki, T., Arnell, N. W., Benito, G., Cogley, J. G., Döll, P., Jiang, T., & Mwakalila, S. S. (2014). Fresh water resources. In C. B. Barros, V. R., D. J. Dokken, K. J. Mach, M. D. Mastrandrea, T. E. Bilir, M. Chatterjee, K. L. Ebi, Y. O. Estrada, R. C. Genova, et al (Eds.), Climate change 2014: Impacts, adaptation, and vulnerability. Part A: Global and sectorial aspects. Contribution of working group II to the fifth assessment report of the intergovernmental panel on climate change; Field; Cambridge University Press: Cambridge, pp. 229–269.

Danga, B. O., Ouma, J. P., Wakindiki, I. I. C., & Bar-Tal, A. (2009). Legume–wheat rotation effects on residual soil moisture, nitrogen and wheat yield in tropical regions. *Advances in Agronomy, 101*, 315–349.

FAO. (2012). Developing a climate-smart agriculture strategy at the country level: Lessons from recent experience; Background paper for the second global conference on agriculture, food security and climate change; food and agriculture organization of the United Nations, Rome, Italy.

FAO. (2015). Statistical yearbook; Word food and agriculture.

Girones, R., Ferrús, M. A., Alonso, J. L., Rodriguez-Manzano, J., Calgua, B., de Abreu, C. A., Hundesa, A., Carratala, A., & Bofill-Mas, S. (2010). Molecular detection of pathogens in water—The pros and cons of molecular techniques. *Water Resource, 44*, 4325–4339. 10.1016/j.watres.2010.06.030.

IPCC. (2013). Summary for policymakers. In T. Stocker, D. Qin, G. K. Plattner, M. Tignor, S. K. Allen, J. Boschung, A. Nauels, Y. Xia, V. Bex & P. M. Midgley (Eds.), *Climate change. The physical science basis. Contribution of working group I to the fifth assessment report of the intergovernmental panel on climate change, stocker*. Cambridge University Press.

Karmakar, R., Das, I., Dutta, D, & Rakshit, A. (2016). Potential effects of climate change on soil properties: A review. *Science International, 4*, 51–73.

Maphosa, Y., & Jideani, V. A. (2017). The role of legumes in human nutrition. In M. C. Hueda (Ed.), *Functional food—Improve health through adequate food.* IntechOpen. 10.5772/intechopen.69127.

Mendelsohn, R. (2014). The impact of climate change on agriculture in Asia. *Journal of Integrative Agriculture, 13*, 660–665. 10.1016/S2095-3119(13)60701-7.

Nuruzzaman, M., Lambers, H., Bolland, M. D. A., Veneklaas, E. J. (2005). Phosphorus benefits of different legume crops to subsequent wheat grown in different soils of Western Australia. *Plant Soil, 271*, 175–187. 10.1007/s11104-004-2386-6.

Oluwajuyitan, T. D., Kadiri, O., Origbemisoye, B. A., Fagbohun, O. F., Ukejeh, R. N., Popoola, O., Olawoye, B. (2021). Cereal grain: A vehicle for improved healthy living. In A. K. Goyal (Ed.), *Cereal grains.* IntechOpen.

Preissel, S., Reckling, M., Schläfke, N., Zander, P. (2015). Magnitude and farm-economic value of grain legume pre-crop benefits in Europe: A review. *Field Crop Research, 175*, 64. 10.1016/j.fcr.2015.01.012.

Sarwar, M. H., Sarwar, M. F., Sarwar, M., Qadri, N. A. & Moghal, S. (2013). The importance of cereals (Poaceae: Gramineae) nutrition in human health: A review. *Journal of Cereals and Oilseeds, 4*(3), 32–35. 10.5897/JCO12.023.

Savova, T. (2002). Effects of weather on growth and development of wintering oats in the spring-summer period. Jubilee session-120 years agricultural science. *Sadovo Scientific Works, 1*(1), 245–248.

Schiermeier, Q. (2014). The parched planet: water on tap. *Nature, 510*, 326–328. 10.1038/510326a.

Singh, B., Aulakh, C., & Walia, S. (2019). Productivity and water use of organic wheat–Chickpea intercropping system under limited moisture conditions in Northwest India. *Renewable Agriculture and Food Systems, 34*(2), 134–143. 10.1017/S1742170517000370.

Tang, X., Bernard, L., Brauman, A., Daufresne, T., Deleporte, P., Desclaux, D., Souche, G., Placella, S. A., & Hinsinger, P. (2014). Increase in microbial biomass and phosphorus availability in the rhizosphere of intercropped cereal and legumes under field conditions. *Soil Biology Biochemistry, 75*, 86–93. 10.1016/j.soilbio.2014.04.001.

Wheeler, T., & Von Braun, J. (2013). Climate change impacts on global food security. *Science, 341*, 508–513. 10.1126/science.1239402.

Contents

Chapter 1
Wheat: High Consumption and Unfulfilled Production

Abstract Global wheat production is expected to be negatively impacted by climate variability and change. In the meantime, the global demand for wheat is expected to increase by 26% by the mid-century to feed the growing population, which will require an annual increase of global wheat production by 2%. Thus, wheat yield performance must be improved to ensure food security for a growing human population amid production constraints due to diminishing resources, land loss and climate change. In this chapter, we discussed two approaches to face the shortage in wheat production and fulfil the existed demand worldwide. Breading for more resilient wheat cultivars to face biotic (viruses, fungi, and insects) and abiotic (heat, drought and salinity) stresses, as well as sustainably increase wheat production are the two suggested approaches to globally reduce wheat production-consumption gap. Globally, breeding programs for resilient wheat cultivars have been implemented and identified genetic resistance against biotic stress in wheat. Furthermore, breeding efforts have been successful to develop resilient wheat cultivars through studying genetic mechanism caused by abiotic stresses and modifying it to develop resistant cultivars. Breeding for improved root system of wheat was also done to improve water and nutrients uptake. With respect to the second approach, namely sustainably increase wheat production, increasing nutrients (nitrogen, phosphorus and potassium) use efficiency should be implemented, as well as reducing the reliance on chemical fertilizer by application of organic fertilizer, wheat rotating with legume crops, and implementing wheat intercropping system with legume crops were also reviewed. Moreover, increasing water use efficiency and water productivity can also conducted to sustainably increase wheat production through improving cultivation method and using more efficient irrigation system. Inclusion of legume crops in wheat-based cropping can lead to increase water use efficiency and water productivity. These practices will increase wheat production and, in the meantime, it will achieve the sustainable use of soil and water resources.

Keywords Biotic and abiotic stresses · Nutrients use efficiency · Water use efficiency · Water productivity · Legumes inclusion in wheat-based cropping system

S. Ouda and A. E. Zohry, *Integration of Legume Crops with Cereal Crops Under Changing Climate*, https://doi.org/10.1007/978-3-031-68102-8_1

Introduction

The recent extensive degradation of land and the exhaustion of water and other natural resources are challenging the sustainability of current food production systems (FAO, 2014). Changes in weather conditions eventually cause extreme climate shifts, which negatively affect the growing crops, and increases insect population (Barlow et al., 2015). Concurrently, the continuous elevation in CO_2 level and the rising in air temperature, as well as changing precipitation patterns and quantities resulting from climate change (Leng & Hall, 2019), in addition to the increasing frequency and severity of extreme climatic events and adverse weather conditions, threaten present global food production (Zampieri et al., 2017).

Wheat is one of the key staple crops for humans in global food security, providing about 20% of total dietary calories and protein needs (Hawkesford et al., 2013). In 2020, the global production of wheat was about 761 million tons produced from a harvested area of around 219 million hectares (FAOSTAT). Globally, humans depend on wheat to obtain 85 and 82% of its required calories and protein, respectively (Chaves et al., 2013). Wheat can grow in various climates and soil types, however temperate regions is the most suitable weather conditions worldwide (Ayed-Slama et al., 2018).

Global wheat production is projected to be negatively impacted by climate change (Barlow et al., 2015). Global warming has resulted in large variations in wheat yield, with expected losses reaching up to 6% for each 1 °C rise in temperature (Miller et al., 2019). On the contrary, climate variability was shown to account for almost 35% of global wheat yield variation (Leng & Hall, 2019). The occurrence of extreme temperature during sensitive stages of wheat, such as flowering, anthesis, and milking stage negatively affect grain size, grain weight, and final wheat yield (Talukder et al., 2014). Furthermore, high temperature reduced spikes production and make it more susceptible to diseases (Martre et al., 2015) and it could increase insect population, which could reduce final wheat yield (Barlow et al., 2015).

It has been reported that climate variability can trigger drought occurrence. It was reported that frequent droughts could increase the risk of wheat yield loss by almost 12% by the end of the twenty-first century (Leng & Hall, 2019). Wheat yield loss under drought is usually due to its effects on the growth of wheat at different growth stages through restriction of photosynthesis (Li et al., 2015). Drought causes low soil water potential, and consequently progressive reduction in leaf relative water content, and water potential (Habash et al., 2014), consequently it reduces plant height, and biomass of wheat (Li et al., 2015). Moreover, Kirigwi et al. (2004) indicated that the most sensitive growth stage to lack of water is the flag leaf stage, followed by the flowering stage. Water shortage in the flag leaf stage reduces the assimilation rate and accelerates the decomposition of assimilation during senescence (Farkas et al., 2020). Mirzaei et al. (2011) reported that drought stress at stages of stem elongation, flowering and grain filling induced 32, 32 and 35% reduction in grain yield, respectively. Whereas, Bauder (2001) indicated that drought stress during maturity resulted in about 10% decrease in wheat yield.

The global demand of wheat is expected to increase by 26% by mid-century (Cui, 2020) to feed the growing population, which will require an annual increase of global wheat production by 2% (Gill et al., 2004). Despite the advancements in wheat production through improved technology (Kukal & Irmak, 2018) and breeding for more resilient cultivars (Xiong et al., 2021), the risk due to climate variability and change has also been elevated (Yu et al., 2018). Wheat yield performance must improve to ensure food security for a growing human population amid production constraints due to diminishing resources, land loss and climate change (Ray et al., 2012). Optimization of agronomic management and sustainable intensification should be considered to overcome this problem (Reynolds et al., 2011). Moreover, climate-smart practices should be implemented to sustainably increase wheat yield, in the meantime conserve natural resources (FAO, 2012). Integration between water-smart, soil-smart and crop-smart practices should be implementing to increase wheat productivity per hectare. Alternatively, increasing the upper limit of genetic yield potential through traditional plant breeding is a successful approach to produce new cultivars with more resilient ability to harsh growing conditions (Reynolds et al., 2011). These new wheat cultivars must have a sufficient level of drought and heat tolerance to avoid significant reductions.

In this chapter, we discussed two approaches to face the shortage in wheat production and fulfil the existed demand worldwide. Breading for more resilient wheat cultivars to combat biotic and abiotic stresses and sustainably increase wheat production are the two suggested approach to globally reduce wheat production-consumption gap.

Breeding for Resilient Wheat Cultivars to Face Biotic and Abiotic Stresses

During growing season, plants are exposed to external stress conditions that could adversely affect its growth, development or productivity. Stresses trigger a wide range of plant responses like altered gene expression, cellular metabolism, and changes in growth rates (Verma et al., 2013). Two primary categories of plant stress are recognized, namely biotic and abiotic stresses. Biotic stress is caused by living organizes, such as viruses, bacteria, fungi, nematodes, insects, and weeds. Whereas, abiotic stress imposed on plants by environment, such as extreme temperatures (cold, frost and heat), drought (water stress), excessive watering (water logging), and salinity (Groppa & Benavides, 2008). Thus, there is an urgency to develop resilient crop cultivars to both stresses categories to ensure food security for the growing populations (Kazi et al., 2013).

Combating Biotic Stress to Minimize Wheat Yield Loss

Biotic stress causes a great threat to wheat production over the globe. In wheat, biotic stress resulted from several living organisms namely fungi (Tembo et al., 2020), virus (Zhang et al., 2017b), insects (Katare et al., 2015), nematodes (Dababat et al., 2015), and weeds (Bertholdsson, 2005). These predators reduce the potential yield performance of wheat cultivars to a massive extent. Yield potential was defined by Evans (1993) as the yield of a well-adapted cultivar as determined by atmospheric carbon dioxide, temperature, and solar radiation in absence of limitations from water, nutrients, weeds, pathogens, and insect pests.

Among biotic stresses, pathogenic fungi represent a significant challenge to wheat production globally (Tembo et al., 2020). The major diseases in wheat involves rust diseases (stripe, stem, and leaf rusts), powdery mildew, and head blight (Figueroa et al., 2017). Historically, yellow rust has caused significant and severe losses in susceptible wheat cultivars worldwide (Wellings, 2011). The major insect-pests attacking wheat are aphid, hessian fly, green bug and borers (Katare et al., 2015).

Bhatta et al. (2019) stated that genetic resistance against biotic stress in wheat is existed in "Synthetic hexaploid wheat" (a cross between *Triticum turgidum* L. × *Aegilops tauschii* Coss.), which have a novel source of genetic diversity for wheat genetic improvement. It possesses genetic diversity for D genome (Bhatta et al., 2018). Jighly et al. (2016) stated that Synthetic hexaploid wheat is resistant to leaf and stem rusts, as well as stripe rust yellow spot (incited by *Pyrenophora tritici-repentis*), Septoria glume blotch (incited by *Parastagonospora nodorum*), crown rot, and multiple pests. Furthermore, it is also resistant to fusarium head blight (incited by *Fusarium graminearum*) and Septoria leaf blotch (incited by *Zymoseptoria tritici*) (Das et al., 2016). The resistant of Synthetic hexaploid wheat to diseases and pests can be used in breeding for resistance to multiple biotic stresses. Additionally, Bhatta et al. (2019) identified several lines having resistance to multiple rusts (Lr, Sr, and Yr) at both the seedling and adult plant stage and their resistance genes can be transferred to bread wheat cultivars.

Overcome Abiotic Stress to Minimize Wheat Yield Loss

The negative effect of extreme temperatures (cold, frost and heat), drought (water stress), excessive watering (water logging), and salinity on wheat can be overcomes by studying the genetic mechanism caused by these stresses and modifying it to develop resistant cultivars (Rollins et al., 2013). At present, wheat genetic variability is available through land races and normal cultivars as conventional crop improvement methods. Accordingly, the germplasm base available for wheat improvement in the form of cultivars is becoming progressively narrow (Talebi, 2011). Doubled haploidy technology is a breeding technique used for wheat breading because it is the quickest method for developing homozygous lines to be used in a breeding program and shorten the breeding cycle, as well as increasing the efficiency of the breeding process (Ferrie & Caswell., 2011). Thus, there is a progressively need for widening the gene pool to combat the emerging biotic and abiotic stresses due to global warming and climate change (Liu et al., 2021).

Breeding for Heat Stress Tolerance in Wheat

Heat stress is one of the major abiotic stresses affecting the growth and yield-related characters of wheat (Sarkar et al., 2021). Mickelbart et al. (2015) indicated that breeding for heat tolerance involves identification of superior genetic stock for progress in heat tolerance, which requires evaluation of wheat genotypes for yield stability under heat stress in different locations. Recurrent selection has effectively exercised to improve grain filling rate and grain size of wheat under heat stress (Gill et al., 2004). Some morphological and physiological traits are the criteria in wheat improvement under heat stress. The morphological traits, such as leaf senescence, grain filling duration and grain yield are used in wheat breeding programs. Premature induction of leaf senescence by heat stress causing chlorophyll destruction and inefficient nutrients recycling initiating enormous yield losses (Vijayalakshmi et al., 2010). The maintenance of photosynthetic capacity and leaf chlorophyll or what is called stay-green character is an indicator of heat tolerance and has been technically used in breeding line selection (Thomas & Ougham, 2014). It was reported that heat stress reduces growing season, especially the maturity period, as well as the grain filling duration of wheat (Ahamed et al., 2010). Therefore, wheat genotypes that mature earlier provide a minimum yield loss comparing to late maturing cultivars and can be considered as compatible to escape heat stress injury and used in breeding programs (Menshawy, 2007). Furthermore, maintaining high grain yield under heat stress is one of the most critical aspects for the improvement of wheat, therefore wheat genotypes possess high and stable yield under heat stress can be used as a donor parent in wheat breeding program for heat tolerance (Al-Otayk, 2010).

Furthermore, physiological traits, such as photosynthetic rate, chlorophyll content, and canopy temperature depression are also widely used to evaluate heat stress tolerance in wheat (Pinto et al., 2010). Photosynthetic rate is adversely affected by heat stress as a result of reduction of photo-assimilates, which leads to the production of reactive oxygen species and oxidative damage (Martins et al., 2016), and deterioration of Photosystem II reaction (Sunil et al., 2019). Heat tolerant plants tend to synthesize various reactive oxygen species scavenging and detoxification systems to create protection against the damaging effects of reactive oxygen species (Hameed et al., 2012). Stay-green associated wheat genotype maintains photosynthesis and grain filling under heat stress through late expression of senescence related genes (Vijayalakshmi et al., 2010), which conserve photosynthetic rate and increase nitrogen remobilization to the maturing wheat grains (Poiroux-Gonord et al., 2013). Stay-green traits were also highly associated with canopy temperature depression, where it was observed that its value was higher under heat stress conditions in stay-green genotypes (Dolferus et al., 2011). Therefore, stay-green trait can be used as selection criteria under heat stress in wheat genotypes (Huang et al., 2012). Wheat biomass and yield are controlled by chlorophyll content under heat stress (Talebi, 2011). Chlorophyll content is a stay-green trait and can be utilized as a selection criterion for heat stress tolerance in wheat (Morales et al., 2020).

The genetic diversity of heat stress tolerance is limited in the elite wheat gene pool (Ogbonnaya et al., 2013). *Aegilops tauschii*, the diploid wild ancestral species and the D genome donor of bread wheat, is a promising genetic resource for breeding (Feldman, 2001), and many *Aegilops tauschii* genes have been introduced to bread wheat (Ogbonnaya et al., 2013). It grows widely in harsh environments in Central Asia and the Middle East and it is expected to carry useful genes for biotic and abiotic stress tolerance (Sohail et al., 2011). Elbashir et al. (2017) selected six tolerant plants from the population with the background of the cultivar 'Norin 61' and established six multiple synthetic derivatives of population of cultivar 'Norin 61' selected as heat stress-tolerant lines and evaluate their agronomic and physiological traits. They obtained two lines acquired heat tolerance, which had higher photosynthesis and stomata conductance, as well as exhibited no reduction in grain yield and biomass under heat stress compared to cultivar 'Norin 61'. Their results also indicated that the multiple synthetic derivatives populations include the diversity of *A. tauschii* is a promising resource to uncover useful quantitative traits derived from this wild species.

Breeding for Drought Stress Tolerance in Wheat

It is projected that global warming and climate change will increase the frequency of drought, and that will adversely affect wheat productivity. Improving drought stress tolerance is a very challenging task for wheat researchers (Sallam et al., 2019). The plant response to water stress involves a complex of genes and processes regulated by interlinked genetic and biochemical mechanisms (Chaves et al., 2009). Drought stress limits photosynthetic rate as a result of stomatal closure to prevent transpirational water loss as a response to drought (Pandey & Shukla, 2015). The genotypes with reduced leaf water loss are believed to be more drought tolerant, and able to conserve their water content (Izanloo et al., 2008). However, closure of stomata, in turn, results in decreased CO_2 availability for photosynthetic carbon metabolism and assimilation, thus limit dry matter accumulation (Centritto et al., 2009). Alteration of plant water relations with decreases in leaf water potential, turgor pressure, and relative water content are also responses of wheat plants to water deficit (Farooq et al., 2009), which represents the driving force for cell expansion (Mondini & Pagnotta, 2015), and growth (Chaves et al., 2009), which are considered the most sensitive processes to drought in wheat (Marček et al., 2019). Moreover, the formation of reactive oxygen species was also reported under water stress conditions, which causing membrane rupture and thereby affect photosynthesis (Osakabe et al., 2014). Thus, breeders and genetic workers should select the wheat cultivars with the ability to sustain photosynthetic apparatus and photochemical efficiency under limited water to restrict reduction in wheat productivity (Sallam et al., 2019).

To optimize partitioning to grain yield, the increase in photosynthetic capacity and the increase in radiation use efficiency can be an important approach to develop new germplasm better adapted to water stress conditions (Reynolds et al., 2009). To improve radiation use efficiency, improvement in interception of light necessary for photosynthesis by adjustment the angle of the leaf and canopy architecture, which has an important role in photosynthesis and the productivity of wheat (Foulkes et al., 2007). It has been suggested that an ideal plant ideotype in cereals is the one that has an increased leaf angle from the top to the basal layers of the canopy, which facilitating the penetration of light into the lower layers of the canopy (Zhu et al., 2010). Wheat cultivars with upright leaves have greater light interception capacity that translates into significant increases in photosynthesis, biomass production, and yield (Choudhury, 2000).

The process of senescence in plants, although is genetically controlled, it also determined by environmental conditions. Plants with delayed senescence or stay-green trait usually have a longer period of grain filling, greater accumulation of assimilated CO_2 throughout the season, and higher yields (Vijayalakshmi et al., 2010). It has been reported that a close relationship between stay-green trait and tolerance to drought is exist (Christopher et al., 2016). Thus, stay-green trait could be useful in designing crops suitable for stressful environmental conditions (Morales et al., 2020).

Breeding for Salinity Stress Tolerance in Wheat

Wheat is known to be sensitive to salinity, which obstructs the growth and development of the plants, leads to low or no productivity under extreme severity of salinity (Nassar et al., 2020). Improving the crop performance by conventional breeding methods, introducing gene markers, and selection of genetically modified genotypes are the basic approaches to produce tolerance against salinity stress in plants (Hasanuzzaman et al., 2011). Selection for tolerant wheat genotypes rely on the use of several tolerance indicators including yield stability, agronomic and physiological traits (Araus et al., 2013). The most wheat yield components that are used to determine the final yield are the spike number per m^2, the number of kernels per spike and the thousand kernel weight, where all reduced under salinity stress leads to a greater influence on the final grain yield (Ayed-Slama et al., 2018). Other studies revealed that biomass, plant height, leaf area, leaf number, tillers number, root length and root biomass could be also use as bases for section criteria for salinity stress in wheat (El-Hendawy et al., 2007).

Some physiological traits such as ability to accumulate ions and osmatic adjustment have been suggested as a salt tolerance trait in plants (Borrelli et al., 2018). Wheat has high ability to exclude Na^+ under salinity conditions (Negrão et al., 2017). The K^+/Na^+ ratio in the root and shoot is an important variable for salinity tolerance in wheat (Hussain et al., 2021). The reduction in plant growth in saline environments could be due to the toxic effects of Na^+ and Cl^- ions on metabolism (Tuna Levent et al., 2008). The ability of young flag leaves and sheaths to maintain low Na^+ and

Cl^- levels reflect the enhanced salt tolerance in some wheat genotypes (Zeng, 2005). Some wheat genotypes preform osmatic adjustment under salinity level and maintain an adequate hydration level to tolerate salt stress.

The relative water content of leaf was found to be a good indicator of salt stress tolerance, which is also highly correlated with wheat yield (Borrelli et al., 2018). Moreover, it has been reported that antioxidant enzyme activities associated with the regulated expression of their encoding genes is existed in salt-tolerant wheat genotypes, resulting in decreased oxidative damage of reactive oxygen species generation (Xie et al., 2019). Oxidative damage of lipids, proteins, and nucleic acids through the alteration of normal cellular metabolism is an important impact of reactive oxygen species (Koca et al., 2007). Furthermore, Liu et al. (2021) reported the important role of cell water integrity in salinity tolerance and propose that the genetic engineering of cell water integrity-related genes using genome editing technology might generate salt-tolerant cultivars for applications in the future.

Breeding for Improved Rooting System in Wheat

Roots are basically important for the plant, where mediation of the water and nutrient uptake required for growth occurs, and they are also a source of chemical and hydraulic signals that modulate shoot growth and physiology (Smith & De Smet, 2012). Root growth response differently under temperature and water stresses (Bishopp & Lynch, 2015). Root growth is diminished under high temperature due to a reduction in the carbon partitioned to it (Batts et al., 1998). Whereas, under moderate water stress, an increased amount of carbon assimilates are sent to the root and root development can be promoted (Smith & De Smet, 2012).

The size of wheat root system provides an opportunity to improve the efficiency of water and nutrients uptake (Wasson et al., 2012). It has been reported that wheat plants with large biomass and more prolific root systems can improve water uptake (Lilley & Kirkegaard, 2011), as well as improved nitrogen uptake in dry environments and deep sandy soils (Palta & Yang, 2014). Water and nutrients uptake by plant roots from a given depth of the soil profile is dependent on root distributional traits, such as the root length and weight (Zhao et al., 2018). Figueroa-Bustos et al. (2018) stated that wheat cultivars with large root systems had 25% more leaf area and biomass than those with small root systems, which apparently reflects high canopy photosynthesis to supply the demand for carbon assimilates to roots. It has been reported that some wheat genotypes that invest significant resources in deep root development have the ability to extract residual moisture under water stress (Lopes & Reynolds, 2010). Root depth and biomass distribution in the soil profile are critical traits for adaptation to water stress and have been prioritized for improving drought resilience in wheat (Gupta et al., 2020).

Root growth was found to have an important role in increasing nitrogen absorption (Carvalho & Foulkes, 2011). Furthermore, a variety of environmental conditions, including soil type and nutrient availability, have a significant impact on root spreading characteristics. Under low soil nitrogen, an increase in the root biomass to total plant biomass ratio was found, which improved the efficiency of nitrogen uptake (Robinson, 2001). Besides, root hairs are an essential feature to be consider for increased nitrogen absorption by active transport, which play a significant role in increasing the surface area of roots (Ehdaie et al., 2003).

Wheat roots has shown a significant level of genetic diversity (Ford et al., 2006). Therefore, more attention should be paid in breeding programs to attain improvement in root distributional traits, namely the root length and weight to improve water and nutrients uptake under stressful growth conditions.

Sustainably Increase Wheat Production

Increasing wheat production in a sustainable manner can be done by several management practices. Taking into consideration the practices of climate-smart agriculture suggested by the FAO organization (2012), wheat production can be increased in a sustainable manner and that can be a pathway to achieve higher wheat productivity by unit area. In the meantime, increasing nutrients and water use efficiency should be taking into account for conserving natural resources and reducing environmental pollution.

Increasing Nutrients Use Efficiency

Fertilizer application is needed to improve wheat yield because wheat, as cereal crop, is characterized by depleting soil nutrients. Therefore, sufficient fertilizers should be added to the soil to prevent its deterioration by replacing nutrients that were uptake by the growing plants or lost during cropping (Liu et al., 2022). In order to increase wheat production, increasing nutrients use efficiency should be achieved by using the proper fertilizer in the right amount and the right time for the plants to maximize crop productivity and reduce pollution (Jan et al., 2007).

However, long-term and excessive use of chemical fertilizers in agricultural production was done to increase crop yield, which is considered to be detrimental to agricultural sustainability (Liu et al., 2022). The right fertilizer amount is dependent on cultivar, growth stage, climate, and management practices, whereas synchronize between crop demand and nutrients supply is necessary to improve nutrients use efficiency (Roberts, 2008).

Nitrogen Fertilizer

Nitrogen is a major macronutrient often limiting plant growth, in which its availability has impacts throughout crop development, namely affecting seedling establishment, tillering, canopy development as well as grain filling and final yield (Hawkesford et al., 2013). Crop responses to nitrogen and its use efficiency is affected by the rate and timing of nitrogen application in relation to different factors, such as plant development, cultivar and climatic conditions (Masaka, 2006). Application of split doses of nitrogen during plant's growing season are known to be effective in increasing nitrogen use efficiency, rather than one single application prior to cultivation (Ooro et al., 2011). Nevertheless, it has been reported that about 50% of applied nitrogen fertilizer remains unavailable to a crop due to temporary immobilization in soil organic matter or due to losses by leaching, erosion nitrification or volatilization (Zafar & Muhammad, 2007).

However, the reliance on nitrogen fertilizer in crop production has an impact on the climate, environment, and public health (UNEP, 2019). Therefore, adopting sustainable nitrogen management resolutions is important to prevent such practices through developing accurate and meaningful nitrogen metrics to identify, monitor, and develop management practices or innovations that reduce nitrogen pollution (Fan et al., 2018). One of these used metrics is nitrogen use efficiency; as it can be used to minimize nutrient losses and the negative impact on surrounding water, air and ecosystems, as well as reducing costs associated with excessive fertilizer inputs (Galloway et al., 2014). Congreves et al. (2021) reviewed and identified numerous approaches used to calculate nitrogen use efficiency to assess its usefulness and shortcomings. They also suggested several factors are currently ignored in traditional nitrogen use efficiency indices, which could improve the conceptualization of nitrogen use efficiency, such as: accounting for a wider range of soil nitrogen forms, considering how plants mediate their response to the soil nitrogen status, including the below-ground/root nitrogen pools, capturing the synchrony between available nitrogen and plant nitrogen demand, blending agronomic performance with ecosystem functioning, and affirming the biological meaning of nitrogen use efficiency.

The productivity and nitrogen use efficiency of wheat is largely affected by many environmental and managerial factors, such as the temperature prevailing during the growing season, irrigation methods, and nitrogen fertilization type (Fan et al., 2018). Other benefit of nitrogen fertilization than increases in wheat grain yield is facilitation the uptake and translocation of Fe and Zn into grains (Shi et al., 2010). Kutman et al. (2011) showed that nearly 80% of the total shoot Zn and 60% of the shoot Fe were harvested with grains at a high nitrogen supply rate. The rate of nutrient absorption in wheat varies with growth stage, variety, growing conditions, and environment. Small amounts of nitrogen are required for seedling viability in the early stages and the mid-tillering stage uses almost half of the nitrogen required (Franzen et al., 2002). According to Shanahan et al. (2008) the application of total nitrogen before or at sowing results in fewer kernels per ear, and number of ears per area, as well as it reduces grain yield in wheat due to poor synchrony. However, over-fertilization could result in low nitrogen use efficiency and groundwater contamination under unfavorable environmental conditions (Lollato et al., 2019).

The root architecture is also thought to be a key factor in nitrogen use efficiency enhancement. A variety of environmental conditions, including soil type and nutrient availability, have a significant impact on root spreading characteristics (Gregory &

Brown, 1989). In wheat, root growth was found to have an important role in increasing nitrogen absorption (Carvalho & Foulkes, 2011). Under low nitrogen circumstances, an increase in the root biomass to total plant biomass ratio was reported to preserve the functional balance between roots and shoot development (Shahzad & Amtmann, 2017). Furthermore, variability on wheat grain yield and protein can also be attributed to differences in cultivar genetic potential, and to crop management (Ladha et al., 2005), as well as plant interactions with environmental variables (Baligar et al., 2001).

Therefore, to increase nitrogen use efficiency, the right amount of nitrogen fertilizer should be synchronized between crop demands. Split applications of nitrogen during the growing season, rather than a single dose, large application prior to planting, are known to be effective in increasing nitrogen use efficiency (Cassman et al., 2002).

Phosphorus Fertilizer

Phosphorus is an essential macro nutrient vital to plant growth and involved in energy transfer, photosynthesis, transformation of sugars and starches, nutrient movement within the plant (Shabnam et al., 2016). Soil phosphorus deficiency is a major constraint in increasing crop yields in many areas of the world (Vance et al., 2003). Relative to nitrogen and potassium, the recovery of phosphorus fertilizers by plants is usually very low due to soils' high capacity to fix phosphorus to soil constituents of little bioavailability (Balemi & Negisho, 2012). As a result, high amounts of phosphorus fertilizers have been applied to soils in order to meet the phosphorus demand of agricultural systems (Tilman et al., 2001). However, crops can usually use only 10–30% of phosphorus in the application year (Syers et al., 2008) and the remaining is fixed as insoluble phosphorus in soils matrix (Nisar, 2000).

Therefore, enhancing the efficiency of phosphorus acquisition and utilization by plants is very important to reduce negative environmental consequences (Clemens et al., 2016). Plants have evolved highly specialized adaptive mechanisms to access soil phosphorus when it is limiting through highly specialized adaptive mechanisms, including an increased root/shoot ratio (Péret et al., 2011), an increase in the number of root hairs, association with arbuscular mycorrhizal fungi (Richardson et al., 2011), synthesis and release of phosphatases and organic acids (Richardson et al., 2009), and enhanced expression of phosphate transporters (Péret et al., 2011). A plant's ability to acquire soil phosphorus could be increased by alterations in root morphology and foraging (Lynch & Brown, 2001). Arbuscular mycorrhizal fungi are widely known to colonize most plant roots and facilitate phosphorus acquisition by roots (Smith & Read, 2010). Deng et al. (2018) reported that arbuscular mycorrhizal fungi colonization of wheat roots was high with low phosphorus application levels and was inhibited by high phosphorus application levels. Synthesis and secretion of phosphatase by wheat roots has been reported to increase phosphorus availability to those roots (George et al., 2008).

It has been reported that wheat crop growth and yield are significantly influenced by phosphorus, where early stages of wheat plants required phosphorus for proper growth and high yield (Havlin et al., 2016). An adequate amount of phosphorus enhances the root growth and seedling establishment of wheat crop (Grant et al., 2005), improves tillering and results in uniformity at heading stage (Grant et al., 2001), and increases the water use efficiency of the plants which ultimately increases the wheat potential of grain yield (Gupta et al., 2001). On the other hand, phosphorus deficiency reduces number of tillers and plant leaf area by producing smaller and lower number of leaves in wheat (Hussain et al., 2008). There is a positive relation between phosphorus and sulfur depends on their rate of application (Hedge & Murthy, 2005). It has been reported that wheat grain and biological yield increases significantly with increasing both phosphorus and sulfur levels (Ibrahim et al., 2012).

Potassium Fertilizer

Potassium is one of the most important elements for growth of plants and plays a vital role in photosynthesis enzyme action, production of protein, carbohydrates and enabling to resist against pest and diseases (El-Shal, 2016). The current literature indicates that at the global level, potassium is as limiting as nitrogen and phosphorus for plant productivity in terrestrial ecosystems (Sardans & Peñuelas, 2015). Potassium plays a vital role in cell metabolism, enhancing enzyme activity and enhanced the plant growth and crop quality (Birch et al., 2008; Haji et al., 2012; Zia-ul-Hassan et al., 2014). Potassium also assists in root system strengthens, improves fruit and vegetable quality, and lessens the effects of crop lodging (Taher & AL-Naser, 2021). It has been reported that potassium helps the plants resists diseases, reduce the effect of drought, frost, and soil salinity (EL Sabagh et al., 2021; Hussain et al., 2020). Furthermore, it is an essential element in the maintenance of osmotic potential and water uptake and had a positive impact on stomata closure which increases tolerance to water stress, water use efficiency and water movement (Nguyen et al., 2002). Potassium application can increase number of fertile tillers in wheat (Mehdi et al., 2001), biomass production (Ehsan et al., 2002), number of grains per spike and wheat grain yield (Evans & Riedell, 2006).

It has been reported that potassium deficiency restricts sustainable crops production (Feng et al., 2014). In wheat, potassium deficiency leads to reductions in both the number and size of leaves produced, resulting in a decrease in the amount of photosynthetic assimilates available for growth and consequently it has negative effects on wheat yield and its quality (Pettigrew, 2008). A decrease in seed size was reported due to insufficient potassium, which in turn affects the grain filling duration (Ashraf et al., 2011).

Several procedures can be used to improve potassium use efficiency. Foliar application of potassium has many advantages over soil fertilization, where it contributes in improving nutrients absorption and transformation efficiency (Gu et al., 2021). The absorption rate and utilization efficiency of nutrients applied to crop leaves are much higher than those of fertilizers applied to soil (Zhang et al., 2010). Compared with soil potassium fertilization, foliar application of potassium fertilizer has the advantages of low cost, no restriction of irrigation conditions and can be combined with insecticides spraying management (Gu et al., 2021). Split application of potassium is another procedure to be done to increase its uptake efficiency than the basal application of one dose (Lu et al., 2014), where it can lower the loss of potassium leaching (Romheld & Kirkby, 2010). The foliar application of potassium can also promote the accumulation of starch in the grain, as well as increase the grain yield of wheat (Liang & Yu, 2004).

Furthermore, many studies have demonstrated that potassium fertilizer efficiency is dependent on optimal nitrogen and phosphorus levels, where its efficiency is lower when potassium fertilizer is applied alone or with phosphorus only, but higher when potassium fertilizer is applied with nitrogen (Burlacu, 2007). Nitrogen aids in the efficient utilization of potassium, phosphorus, and other nutrients by plants (Brady & Weil, 1984). Nevertheless, nitrogen and phosphorus use efficiency, as well as productivity and quality of agriculture produce, could all benefit from increased potassium fertilization (Niu et al., 2013).

Reducing the Reliance on Chemical Fertilizer in Wheat Production

Application of Organic Fertilizer

The unwise use of chemical fertilizers is very hazardous to the environment. Fertilization strategies that provide food security and environmental sustainability, while reducing fertilizer application are urgently required. Soil quality degradation, such as soil nutrients imbalance, soil acidity and soil pollution, have been widely reported because of long-term or excessive application of chemical fertilizers (Yang et al., 2018). Nevertheless, organic fertilizer sources not just supply the required plant nutrients, but also positively affect plant growth and development, soil water holding capacity, as well as improving the fertility of the soil (Wick et al., 2017). Application of organic fertilizer could represents an agricultural management practice that obtain a more sustainable wheat production and improve grain quality, as well as its nutritional value as an alternative to mineral nitrogen fertilizers (Bezabeh et al., 2022).

Organic fertilizers are characterized by abundant nutrients and organic materials, with slow nutrient release (Liu et al., 2022) and it can improve soil fertility and plant's fertilizer use efficiency (Abd-Elrahman et al., 2022). Therefore, organic fertilizer can be partially substituting chemical fertilizers in crops production (Cen et al., 2020). Yang et al. (2020) reported that using organic fertilizer to partially substituting chemical fertilizers provides the required nutrients that increases wheat yield and nutrient utilization and absorption. A reduction in the soil bulk density was reported when organic fertilizer was applied (Blanco-Canqui et al., 2015). Dai et al. (2021) reported an improvement in soil physicochemical properties, enzymatic activity and the micro-ecological environment, when organic fertilizer substitution was used. A study by Yang et al. (2021) revealed that application of organic fertilizer to wheat released nitrogen slowly in the early stage and showed a significant residual effect, whereas when urea was applied, nitrogen was quickly released in the first two months. Substitution of organic fertilizer with 30–50% urea, increased wheat yield by 4–9%. Furthermore, nitrogen leaching to the 60–100 cm soil layer was significantly reduced. Hence, they determined that this strategy is the best for simultaneously increasing wheat production and reducing nitrogen leaching.

Application of animal manure, as a source of organic fertilizer, is very important for soil fertility, and enhances the crops production. Gupta et al. (2012) reported that application of animal manure improves the soil physical properties; increase soil infiltration rate and water holding capacity, in addition to being environment friendly. Aula et al. (2019) reported an increase in wheat yield by 66% in the second year after manure application, with no change in yields in the third and fourth year. Additionally, higher value of soil organic matter was also observed. Akhtar et al. (2020) Mixed phosphate fertilizer with farmyard manure to increase phosphorus availability and improving the productivity of wheat grown in alkaline calcareous soils. A higher values of phosphorus ions concentration in the soil solution and exchangeable phosphorus were found resulted in higher wheat grain yield and phosphorus use efficiency.

Poultry manure is another source of organic fertilizer, characterized by being cheap and excellent source of phosphorus among all organic fertilizer sources. It contains around 3% nitrogen, 3% phosphorus and 1% potassium (Babu & Ibraham, 2006).

Composts from plant residues are inexpensive organic fertilizers that have traditionally been used in many areas. However, because of slow decomposition processes and thus low available nutrient levels, applying composts as fertilizer requires large amounts of biomass (Bezabeh et al., 2022). Kabato et al. (2022) stated that the integrated application of chemical fertilizer and compost significantly increased soil organic carbon, total nitrogen, and available phosphorus, in addition to producing higher wheat grain yield than the obtained under application of chemical fertilizer only. Similar results were obtained by Mohamed et al. (2019), where application of compost with chemical fertilizer positively affected wheat yield.

On the other hand, it has been reported that vermicompost application enhances crop growth and wheat yield compared to other non-composted organic wastes (Kizilkaya et al., 2012) and mineral fertilizers (Joshi et al., 2013). Vermicompost is finely-divided mature peat-like materials which are produced by a non-thermophilic process involving interactions between earthworms and microorganisms leading to bioxidation and stabilization of organic material (Aira et al., 2000). Kutman et al. (2011) stated that vermicompost improved the biomass and yield of wheat compared to mineral fertilizer and animal manure. A previous study revealed that vermicomposting amendments could improve soil quality and production by enhancing the chemical traits of the soil, augmenting the amount of plant-available nutrients, improving soil biological activities, and boosting crop yield (Goswami et al., 2017). In addition, other studies reported that wheat yield was positively affected by the application of vermicompost in combination with other organic or mineral fertilizers (Aslam et al., 2019; Erdal & Ekinci, 2020; Joshi et al., 2013).

Wheat Rotating with Legume Crops

Wheat rotating with legume crops can increase wheat yield in a more sustainable way as a result of some of the nitrogen added to the system by the legume crops is available for the following crops, as a result the crops yield in the rotation can be increases (Bezabeh et al., 2022). Mesfin et al. (2023) stated that wheat rotation with faba bean had a higher grain yield and nutrient concentration, than continuous wheat production. Kirkegaard et al. (2008) also found an increase in wheat yield from 20 to 36% in a faba bean–wheat rotation compared to its value in barley–wheat rotation. Soil nitrogen is enriched by various grain legumes through biological nitrogen fixation which subsequently enhances wheat yields (Fustec et al., 2010). Furthermore, other benefits occur from rotating faba been with wheat includes reduction of wheat root rot incidence, and wheat leaf disease and pests (Bezabeh et al., 2022). Legume–cereal rotations are also known to reduce the demand for labor for weed control in subsequent cereal crops (Vereijken & Kloen, 1994). In addition, legumes reduce soil erosion (Lawson et al., 2007) and enhance yield stability and resilience (Suso & Maalouf, 2010).

Introduction of legume crop, such as chickpea, field bean, faba beans in wheat-based cropping is a viable strategy for the reduction of inorganic fertilizer use (Danga et al., 2009). It also contributing to reduced CO_2 emissions (Nieder & Benbi, 2008) and carbon footprints of agricultural products (Gan et al., 2011). Several studies reported savings between 100 and 200 kg/ha in the amount of nitrogen fertilizers applied to cereal crops following faba bean (Nuruzzaman et al., 2005; Preissel et al., 2015).

Wheat Intercropping Systems with Legume Crops

The system of legume intercropping with cereals is the most used in the sustainable agriculture around the world. The advantages of intercropping with legumes consist mainly of the complementary use of nitrogen sources by the species used in the intercropping (Dhima et al., 2014), where it plays a key role in promoting ecosystem efficiency (Vandermeer et al., 1998), especially organic farming (Bedoussac et al., 2015). Legumes are fixing nitrogen in the roots by rhizobium through symbiotic nitrogen fixation (Pelzer et al., 2012). Intercropped legumes have proved to producing substantially higher yields than a sole crop expressed as a land equivalent ratio higher than 1.0 (Yu et al., 2015). It also has a facilitative function in mitigating stressful conditions through plant root interactions and mobilizing limited or unavailable nutrients, such as phosphorus in harsh environmental conditions (Latati et al., 2016). Furthermore, it has a key role in soil organism diversity in the rhizosphere (Tang et al., 2014). Intercropping systems of cereal with legumes not only allow the optimal use of soil and atmospheric nitrogen sources, but also improve soil conservation and increase yield and protein concentration in the cereal grains, especially under low soil nitrogen levels. Moreover, it is also helping in weed control and reduce diseases and pest occurrence (Bedoussac et al., 2015).

Dvořák et al. (2022) stated that intercropped wheat with legume crops achieved higher grain yield compared to wheat pure stand, especially under organic cultivation. Reganold and Wachter (2016) found that, in organic production, wheat yield averages are 8 to 25% lower compared with traditional agriculture with lower cost. Furthermore, intercropping pea with wheat proved to reduce the amount of nitrogen fertilization and production costs, while ensuring production with less pollution from the mineral nitrogen (Jensen et al., 2020). Carlsson et al. (2017) reported that, in pea intercropping system with wheat, wheat is more competitive in its use of mineral nitrogen, impulses pea to fix more atmospheric nitrogen by symbiosis to meet its needs, which hinders the development of weeds due to the lack of nitrogen resources. The amount of nitrogen fixed by intercropped pea is higher than that of sole pea due to competition with wheat under intercropping (Hauggaard-Nielsen et al., 2001). Fan et al. (2006) reported that proportion of total nitrogen derived from fixation increase in wheat-faba bean intercropping system, compared to maize-faba bean intercropping system without nitrogen fertilization condition. Furthermore, Chapagain and Riseman (2014) indicated that intercropping wheat with faba bean attained higher land equivalent ratio and total land outputs, with increased land productivity, as well as improved wheat biomass, and grain protein content, compared to monoculture cultivation of wheat.

Increasing Water Use Efficiency and Water Productivity

Irrigation is essential for ensuring the production of food, and it is expected to rise in the future due to high population growth rate and raising global temperature. Producing more food from less water should be the path to be pursued under the prevailing water scarcity conditions. Crop yield increases have been attained at the expense of greater amounts of irrigation water used, therefore the challenge is to reduce the rate of water use without a concomitant reduction in yield (Mei et al., 2013). Water use efficiency describes the intrinsic trade-off between carbon fixation and water loss, because water evaporates whenever stomata open for CO_2 acquisition for photosynthesis (Bramley et al., 2013). Water use efficiency is defined as crop yield per unit of water used (Chai et al., 2014).

Alternatively, crop water productivity is a quantitative term used to define the relationship between crop produced and the amount of water involved in crop production (Igbadun et al., 2006). Water productivity decreases with the increase in the applied irrigation amounts, as yield gain is less than the proportional increase in the applied irrigation water (Malve et al., 2017). Increasing crop water productivity is critical to match the decreasing availability of water and land for food production, where information about water use and efficiency are required to improve water productivity (Brauman et al., 2013). This information will enable to produce 'more crop per drop', thus helping increase food production under the anticipated decrease in land and water allocations (Molden et al., 2007).

Several strategies have been studied to decrease water consumption and enhance water productivity. Malve et al. (2017) reported that water productivity can be increased by increasing the yield of crop without using more water, or by reducing water while maintaining or increasing yield. The following practices can be used to enhance water use efficiency and water productivity:

Raised Beds Cultivation

It is well known that surface irrigation is characterized by the wasteful use of water resources. To improve the efficiency of surface irrigation system, cultivation on raised beds has been proved to be a method with many advantages. In raised beds cultivation, irrigation water is sent through the ditches dug between the beds to provide water to plants, where water seep into raised beds through capillary action, thus increase water use efficiency (Ahmad et al., 2009). Irrigation of raised beds permits more efficient use of water as compared to the basin or border irrigation (Hassan et al., 2005). Raised beds cultivation improves the method of weed control, and crop stand, as well as facilitates mechanical practices during the crop growing season (Miah et al., 2015) and requires less labor (Fischer et al., 2005). Raised beds planting also created better soil physical environment throughout the crop growth period, which led to higher crop productivity (Aggarwal & Goswami, 2003). It improves nutrients uptake due to less leaching loss of nutrients, as well as enhances the availability of sufficient moisture for mineralization of the applied nutrients (Idnani & Kumar, 2013). Other studies on raised beds cultivation showed that it reduces seed mortality rates, and improves soil quality (Limon-Ortega et al., 2002), which led to enhanced root growth, and higher yield (Dey et al., 2015).

Raised beds cultivation of wheat proved to increase water use efficiency and water productivity. Sing et al. (2010) found lower water consumption and higher wheat yield under raised beds planting than that under conventional flat beds planting due to decrease in irrigation amount. Furthermore, Majeed et al. (2015) indicated that raised beds planting of wheat, not only saved water but also improved fertilizer use efficiency and increase grain yield by 15%, compared to flat planting, thus increase water productivity. Hussain et al. (2018) reported that, in wheat cultivated on raised beds, 30–35% of the applied irrigation water was saved, with 13% increase in grain yield, compared to flat border irrigation method. Similarly, the average irrigation duration decreased by 36% in the case of raised beds irrigation method (Hassan et al., 2005) and fertilizer use efficiency improves by 15% (Majeed et al., 2015).

Irrigation System

Drip irrigation is the most efficient among all the irrigation methods, particularly in areas with water scarcity. It involves application of water drop by drop through closed network of plastic pipes at frequent intervals near the root zone for consumptive use of the crop (Suryavanshi & Buttar, 2018). It helps in reducing the over-exploitation of groundwater that partly occurs because of inefficient use of water under surface method of irrigation (Damor et al., 2021). It has been reported that surface irrigation help achieve yield gains of up to 100% and water savings up to 40–80% (Burney et al., 2010). Environmental problems associated with the surface method of irrigation like water logging and salinity are minimize under drip irrigation, as well as it achieved saving in irrigation water, thus increasing water productivity, increasing water use efficiency and saving in associated fertilizer, pesticide, and labor, compared to surface irrigation systems (Lodhi et al., 2014). Vanitha (2008) explained that the enhancement in water use efficiency under drip irrigation was due to enhanced grain yield, as well as lower water consumption and loss of water through evaporation than other irrigation methods. Furthermore, scheduling of irrigation through drip irrigation, provides the required amount of water at critical growth stages of plants, which matches the crop evapotranspiration rate, resulting in high water use efficiency (Mehriya et al., 2020). It has been reported that drip irrigation recorded the highest water productivity over surface basin irrigation method (Malve et al., 2017).

Drip irrigation saves more than 20% of irrigation water, as compare to surface irrigation in wheat crop (Kumari et al., 2014). Rao et al. (2016) reported that drip irrigated wheat gave higher water use efficiency and saved more water than border irrigation method, where water productivity was significantly higher than sprinkler irrigated wheat and wheat cultivated in basin (Suryavanshi & Buttar, 2018). Similarly, Chouhan et al. (2015) revealed that drip irrigation produced higher water productivity in wheat, compared to its value obtained under border irrigation due to the lower water losses in deep percolation, and in evaporation from the soil. Whereas, Bhunia et al. (2015) stated that that drip irrigation levels at 60% of crop evapotranspiration attained higher water use efficiency in wheat as compared to the obtained under surface irrigation.

Incorporation of Organic Fertilizer with Chemical Fertilizer

Organic fertilizer incorporation in the soil enhances the organic matter content leading towards moisture conservation and plant nutrient availability (Hammad et al., 2017). Application of organic fertilizer have showed an increase the soil water-holding capacity (Wang et al., 2011), improving yield and consequently water use efficiency. Liu et al. (2013) found that the increase in soil water holding capacity by organic fertilizer application is due to increasing the percentage of macro-aggregates (> 0.25 mm) and soil nutrients, which ultimately improved yield and water use efficiency. Wang et al. (2011) stated that the significant increases in water use efficiency under organic fertilizer application was not a function of higher water uptake, but a function of soil fertility improvements. Furthermore, it has been reported that application of organic fertilizer enabled plants to use water more efficiently under water deficiency (Wang et al., 2017).

Application of organic fertilizer resulted in significant increase and improvement in the water productivity of wheat under water stress (Bodruzzaman et al., 2010). Shah et al. (2013) confirmed that there was a significant increase in the productivity of wheat crop when organic fertilizer was applied in combination with mineral fertilizers. Furthermore, Alkhuzai (2023) indicated that application of organic fertilization in addition to mineral fertilization allowed to save 25% of required irrigation water in wheat, which increased wheat productivity by 29–56% in two growing seasons. Zhang et al. (2017a) reported that soil water content, grain yield and water use efficiency was higher by 4, 12 and 13%, respectively when manure was applied to wheat, compared to without manure application.

Rotation with Legume Crops

The main benefits that wheat receives from rotating with a legume crop is the increase in yield as a result of the organic matter leftover in the soil. Kröbel et al. (2014) reported that inclusion of a legume crop in a rotation with wheat had a positive effect on water use efficiency of wheat. Rotation with a legume crop positively affects water use efficiency of spring wheat not only with respect to water use of the previous crop (Campbell et al., 2007), but also with respect to plant–plant interactions (Kröbel et al., 2014). Furthermore, Seymour et al. (2012) found that including a legume crop in the rotation increased wheat yield and improve its water use efficiency.

Intercropping with Legume Crops

Intercropping is known by its ability to increase soil water conservation, reduces run-off, increases the use of available soil water, increases crop yield for the entire systems, increases the yield per unit of water applied, and increases the yield of crops grown the following year in the rotation, as well as increases water use efficiency (Blum 2011; Chen et al., 2018; Fan et al., 2013; Yang et al., 2011). It has been reported that intercropping can increase water use efficiency by 4% to 99% compared to its value under sole crops, especially when water supply is not limited (Mao et al., 2012; Stomph et al., 2020; Tan et al., 2020). The root system can also play an important role in water acquisition in intercrops, where species mixtures have been shown to stimulate increased root length density leading to decreased water loss through evaporation (Pankou et al., 2021).

Bouras et al. (2023) indicated that intercropping chickpea with wheat promotes an improvement in growth and yield quality as a result of optimization of water and nitrogen use under low and moderate nitrogen application. Similarly, Singh et al.

(2019) reported that chickpea intercropping system with wheat gave higher water use efficiency than its value in sole wheat. Moreover, intercropping pea with wheat uses the water resources of the environment more efficiently and can be used in dry land conditions for higher yield (Pankou et al., 2021).

Conclusion

In this chapter, we reviewed the global efforts to increase wheat production to help in feeding the growing population. Breeding programs were implemented to produce resilient wheat cultivars that can combat the biotic and abiotic stresses. Additionally, several practices were implemented to sustainably increase wheat production. Of these practices, inclusion of legume crops within the wheat-based system, namely rotating with legume crops, or wheat intercropping with legume crop proved to improve nutrients use efficiency, as well as water use efficiency and water productivity. These practices can contribute in increasing wheat production and, in the meantime, it will achieve the sustainable use of soil and water resources.

References

Abd-Elrahman, Sh. H., Saudy, H. S., Abd El-Fattah, D. A., & Hashem, F. A. (2022). Effect of irrigation water and organic fertilizer on reducing nitrate accumulation and boosting lettuce productivity. *Journal of Soil Science and Plant Nutrition, 22*, 2144–2155.

Aggarwal, P., & Goswami, B. (2003). Bed planting system for increasing water use efficiency of Wheat (T. Aestibum) grown in Inseptisol. *Indian Journal of Agricultural Sciences, 73*, 422–425.

Ahamed, K. U., Nahar, K., & Fujita, M. (2010). Sowing date mediated heat stress affects the leaf growth and dry matter partitioning in some spring wheat (Triticum aestivum L.) cultivars. *The IIOAB Journal, 1*(3), 1–9.

Ahmad, I. M., Qubal, B., Ahmad, G., & Shah, N. H. (2009). Maize yield, plant tissue and residual soil N as affected by nitrogen management and tillage system. *Journal of Agricultural and Biological Sciences, 1*(1), 19–29.

Aira, M., Monroy, F., Dominguez, J., & Mato, S. (2000). How earthworm density affects microbial biomass and activity in pig manure. *European Journal of Soil Biology, 38*, 7–10.

Akhtar, M., Ikram, W., Mahmood, T., Yousaf, S., Waqas Gillani, S., & Ejaz, A. (2020). Improving soil phosphorus supply and wheat yield with manure-amended phosphate fertilizer. *Experimental Agriculture, 56*(1), 48–58. https://doi.org/10.1017/S0014479719000097

Alkhuzai, K. A. H. (2023). The use of organic waste in sustainable water management under the conditions of dry areas: Field and modeling study. *Arabian Journal for Science and Engineering, 48*, 5267–5287. https://doi.org/10.1007/s13369-022-07370-8

Al-Otayk, S. M. (2010). Performance of yield and stability of wheat genotypes under high stress environments of the central region of Saudi Arabia. *The Meteorology, Environment and Arid Land Agriculture, 21*(1), 81–92.

Araus, J. L., Cabrera-Bosquet, L., Serret, M. D., Bort, J., & NietoTaladriz, M. T. (2013). Comparative performance of 13C, 18O and 15N for phenotyping durum wheat adaptation to a dryland environment. *Functional Plant Biology, 40*, 595–608.

Ashraf, M. A., Ahmad, M. S. A., Ashraf, M., Al-Qurainy, F., & Ashraf, M. Y., et al. (2011). Alleviation of waterlogging stress in upland cotton (Gossypium hirsutum L.) by exogenous application of potassium in soil and as a foliar spray. *Crop Pasture Science, 62*(1), 25–38.

Aslam, Z., Bashir, S., Hassan, W., Bellitürk, K., Ahmad, N., Niazi, N. K., Khan, A., Khan, M. I., Chen, Z., & Maitah, M. (2019). Unveiling the efficiency of vermicompost derived from different biowastes on wheat (Triticum aestivum L.) plant growth and soil health. *Agronomy, 9*(12), 791.

Aula, L., Omara, P., Dhillon, J. S., Fornah, A., & Raun, W. R. (2019) Influence of applied cattle manure on winter wheat (Triticum aestivum L.) grain yield, soil pH and soil organic carbon. *Communications in Soil Science and Plant Analysis, 50*(16), 2056–2064. https://doi.org/10.1080/00103624.2019.1648664

Ayed-Slama, O., Bouhaouel, I., Chamekh, Z., Trifa, Y., Sahli, A., Aissa, N. B., & Slim-Amara, H. (2018). Genetic variation of salt-stressed durum wheat (Triticum turgidum subsp. durum Desf.) genotypes under field conditions and gynogenetic capacity. *Journal of Genetic Engineering and Biotechnology, 16*, 161–167.

Babu, S. D., & Ibraham, T. (2006). Sustainable nutrient management in rice cropping system. *Agrobios Newsletter, 4*(12), 48–49.

Balemi, T., & Negisho, K. (2012). Management of soil phosphorus and plant adaptation mechanisms to phosphorus stress for sustainable crop production: A review. *Journal of Soil Science and Plant Nutrition, 12*, 547–561.

Baligar, V. C., Fageria, N. K., & He, Z. L. (2001). Nutrient use efficiency in crop plants. *Journal of Communication Soil Science, 32*, 921–950.

Barlow, K. M., Christy, B. P., O'leary, G. J., Riffkin, P. A., & Nuttall, J. G. (2015). Simulating the impact of extreme heat and frost events on wheat crop production: A review. *Field Crops Research, 171*, 109–119.

Batts, G. R., Ellis, R. H., Morison, J. I. L., Nkemka, P. N., Gregory, P. J., & Hadley, P. (1998). Yield and partitioning in crops of contrasting cultivars of winter wheat in response to CO_2 and temperature in field studies using temperature gradient tunnels. *Journal of Agricultural Science, 130*, 17–27.

Bauder, J. (2001). Irrigating with limited water supplies. Montana State University Communications their relationship with growth and yield in wheat cultivars at different growth stages. *Journal of Agronomy, 86*, 1437–1439.

Bedoussac, L., Journet, E.-P., Hauggaard-Nielsen, H., Naudin, C., Corre-Hellou, G., Jensen, E. S., Prieur, L., & Justes, E. (2015). Ecological principles underlying the increase of productivity achieved by cereal-grain legume intercrops in organic farming. A review. *Agronomy of Sustainable Development, 35*, 911–935. https://doi.org/10.1007/s13593-014-0277-7

Bertholdsson, N. O. (2005). Early vigor and Allelopathy—Two useful traits for enhancing barley and wheat competitiveness against weeds. *Weed Research, 45*, 94–102.

Bezabeh, M. W., Haile, M., Sogn, T. A., & Eich-Greatorex, S. (2022). Wheat (*Triticum aestivum*) production and grain quality resulting from compost application and rotation with faba bean. *Journal of Agriculture and Food Research, 10*, 100425.

Bhatta, M., Morgounov, A., Belamkar, V., Poland, J., & Baenziger, P. S. (2018). Unlocking the novel genetic diversity and population structure of synthetic hexaploid wheat. *BMC Genomics, 19*, 591. https://doi.org/10.1186/s12864-018-4969-2

Bhatta, M., Morgounov, A., Belamkar, V., Wegulo, S. N., Dababat, A. A., Erginbas-Orakci, G., Bouhssini, M. E., Gautam, P., Poland, J., Akci, N., Demir, L., Wanyera, R., & Baenziger, P. S. (2019). Genome-wide association study for multiple biotic stress resistance in synthetic hexaploid wheat. *International Journal of Molecular Sciences, 20*(15), 3667. https://doi.org/10.3390/ijms20153667

Bhunia, S. R., Verma, I. M., Arif, M., Gochar, R., & Sharma, N. C. (2015). Effect of crop geometry, drip irrigation and bio-regulator on growth, yield and water use efficiency of wheat (Triticum aestivum L.). *International Journal of Agricultural Sciences, 11*, 45–49.

Birch, C. J., McLean, G., & Sawers, A. (2008). Analysis of high yielding maize production–A study based on a commercial crop. *Australian Journal of Experimental Agriculture, 48*(3), 296–304.

Bishopp, A., & Lynch, J. P. (2015). The hidden half of crop yields. *Nature Plants, 1*, 15117. https://doi.org/10.1038/nplants.2015.117

Blanco-Canqui, H., Hergert, G. W., & Nielsen, R. A. (2015). Cattle manure application reduces soil compactibility and increases water retention after 71 years. *Soil Science Society of America Journal, 79*, 212–223. https://doi.org/10.2136/sssaj2014.06.0252

Blum, A. (2011). Plant breeding for water-limited environment. Springer. ISBN 978-1-4419-7490-7.

Bodruzzaman, B., Meisner, C. A., Sadat, M. A., & Hossain, M. I. (2010). Long term effects of applied organic manures and inorganic fertilizers on yield and soil fertility in a wheat-rice cropping pattern (pp. 142–145). World Congress of Soil Science.

Borrelli, V. M. G., Brambilla, V., Rogowsky, P., Marocco, A., & Lanubile, A. (2018). The enhancement of plant disease resistance using CRISPR/Cas9 technology. *Frontier in Plant Sciences, 9*, 1245.

Bouras, F.-Z., Hadjout, S., Haddad, B., Malek, A., Aitmoumene, S., Gueboub, F., Metrah, L., Zemmouri, B., Kherif, O., Rebouh, N.-Y., et al. (2023). The effect of nitrogen supply on water and nitrogen use efficiency by wheat-chickpea intercropping system under rain-fed Mediterranean conditions. *Agriculture, 13*, 338. https://doi.org/10.3390/agriculture13020338

Brady, N. C., & Weil, R. R. (1984). Soil reaction: Acidity and alkalinity. In *The nature and properties of soils* (pp. 189–22230). Macmillan Publishing Co.

Bramley, H., Turner, N. C., & Siddique, K. H. M. (2013). Water use efficiency. In C. Kole (Ed.), *Genomics and breeding for climate-resilient crops* (vol. 2, pp. 225–268). https://doi.org/10.1007/978-3-642-37048-9_6

Brauman, K. A., Siebert, S., & Foley, J. A. (2013). Improvements in crop water productivity increase water sustainability and food security—A global analysis. *Environmental Research Letters, 8*(2), 1–7.

Burlacu, G. (2007). Research results on the rational use of fertilizers. Analele INCDA Fundulea LXXV, Volum jubiliar, pp. 287–307.

Burney, J., Woltering, L., Burke, M., Naylor, R., & Pasternak, D. (2010). Solar-powered drip irrigation enhances food security in the Sudano-Sahel. *Proceedings of the National Academy of Sciences USA, 07*(5), 1848–1853. https://doi.org/10.1073/pnas.0909678107

Campbell, C. A., VandenBygaart, A. J., Zentner, R. P., McConkey, B. G., Smith, W., Lemke, R., Grant, B., & Jefferson, P. G. (2007). Quantifying carbon sequestration in a minimum tillage crop rotation study in semiarid southwestern Saskatchewan Can. *Journal of Soil Science, 87*, 235–250.

Carlsson, G., Bedoussac, L., Cupina, B., Djordjevic, V., Gaudio, N., Jensen, E.-S., Jeuffroy, M.-H., Journet, E.-P., Justes, E., & Mikic, A. (2017) Does a mixture of pea varieties with different leaf morphology improve crop performance? In *Proceedings of the international conference on advances in grain legume cultivation and use*, Novi Sad, Serbia, 27–28 September 2017.

Carvalho, P., & Foulkes, M. J. (2011). Roots and the uptake of water and nutrients. In R. A. Meyers (Ed.), *Encyclopedia of sustainability science and technology* (pp. 1390–2140). Springer.

Cassman, K. G., Dobermann, A., & Walters, D. T. (2002). Agroecosystems, nitrogen use efficiency, and nitrogen management. *AMBIO, 31*, 132–140.

Cen, Y., Guo, L. Y., Liu, M. Z., Gu, X., Li, C. H., & Jiang, G. M. (2020). Using organic fertilizers to increase crop yield, economic growth, and soil quality in a temperate farmland. *Peer Journal, 8*. Article e9668.

Centritto, M., Lauteri, M., Monteverdi, M. C., & Serraj, R. (2009). Leaf gas exchange, carbon isotope discrimination, and grain yield in contrasting rice genotypes subjected to water deficits during the reproductive stage. *Journal of Experimental Botany, 60*, 2325–2339. https://doi.org/10.1093/jxb/erp123

Chai, Q., Gan, Y., Turner, N. C., Zhang, R. Z., Yang, C., Niu, Y., & Siddique, K. H. M. (2014). Water-saving innovations in Chinese agriculture. *Advances in Agronomy, 126*, 147–197. https://doi.org/10.1016/B978-0-12-800132-5.00002-X

Chapagain, T., & Riseman, A. (2014). Intercropping wheat and beans: Effects on agronomic performance and land productivity. *Crop Science*. https://doi.org/10.2135/cropsci2013.12.0834

Chaves, M. M., Flexas, J., & Pinheiro, C. (2009). Photosynthesis under drought and salt stress: Regulation mechanisms from whole plant to cell. *Annals of Botany, 103*, 551–560.

Chaves, M. S., Martinelli, J. A., Wesp Guterres, C., Graichen, F. A. S., Brammer, S. P., Scagliusi, S. M., & Consoli, L. (2013). The importance for food security of maintaining rust resistance in wheat. *Food Security, 5*(2), 157–176. https://doi.org/10.1007/s12571-013-0248-x

Chen, G. X., Kong, Y., Gan, R., Zhang, F., Feng, A., Yu, C., Zhao, S., & Wan, Q. (2018). Chai Enhancing the systems productivity and water use efficiency through coordinated soil water sharing and compensation in strip intercropping. *Science and Reports, 8*, 10494.

Choudhury, B. J. (2000). A sensitivity analysis of the radiation use efficiency for gross photosynthesis and net carbon accumulation by wheat. *Agricultural and Forest Meteorology, 101*, 217–234.

Chouhan, S., Awasthi, M., & Nema, R. (2015). Studies on water productivity and yields responses of wheat based on drip irrigation systems in clay loam soil. *Indian Journal of Science and Technology, 8*, 650–654.

Christopher, J. T., Christopher, M. J., Borrell, A. K., Fletcher, S., & Chenu, K. (2016). Stay-green traits to improve wheat adaptation in well-watered and water-limited environments. *Journal of Experimental Botany, 67*, 5159–5172.

Clemens, C., van de Wiel, M., van der Linden, C. G., & Scholten, O. E. (2016). Improving phosphorus use efficiency in agriculture: Opportunities for breeding. *Euphytica, 207*, 1–22. https://doi.org/10.1007/s10681-015-1572-3

Congreves, K. A., Otchere, O., Ferland, D., Farzadfar, S., Williams, S., & Arcand, M. M. (2021). Nitrogen use efficiency definitions of today and tomorrow. *Frontiers in Plant Science, 12*, 637108. https://doi.org/10.3389/fpls.2021.637108

Cui, X. (2020). Beyond yield response: Weather shocks and crop abandonment. *Journal of the Association of Environmental and Resource Economists, 7*(5), 901–932. https://doi.org/10.1086/709859

Dababat, A. A., Muminjanov, H., & Smiley, R. W. (2015). Nematodes of small grain cereals: Current status and research. FAO.

Dai X. L., Song, D. L., Zhou, W., Liu, G. R., Liang, G. Q., He, P., Sun, G., Yuan, F. S., Liu, Z. B., Yao, Y. K., & Cui, J. (2021). Partial substitution of chemical nitrogen with organic nitrogen improves rice yield, soil biochemical indictors and microbial composition in a double rice cropping system in south China. *Soil and Tillage Research, 205*, Article 104753.

Damor, N. N., Solanki, R. M., Hadiyal, J. G., & Dharaiya, B. K. (2021). Effect of drip irrigation schedules and fertigation levels on growth parameters and yields of wheat (Triticum aestivum L.). *Journal of Pharmacognosy and Phytochemistry, 10*(2), 1546–1550.

Danga, B. O., Ouma, J. P., Wakindiki, I. I. C., & Bar-Tal, A. (2009). Legume–wheat rotation effects on residual soil moisture, nitrogen and wheat yield in tropical regions. *Advances in Agronomy, 101*, 315–349.

Das, M. K., Bai, G., Mujeeb-Kazi, A., & Rajaram, S. (2016). Genetic diversity among synthetic hexaploid wheat accessions (Triticum aestivum) with resistance to several fungal diseases. *Genetic Resources and Crop Evolution, 63*, 1285–1296. https://doi.org/10.1007/s10722-015-0312-9

Deng, Y., Teng, W., Tong, Y.-P., Chen, X.-P., & Zou, C.-Q. (2018). Phosphorus efficiency mechanisms of two wheat cultivars as affected by a range of phosphorus levels in the field. *Frontiers in Plant Science, 9*, 1614. https://doi.org/10.3389/fpls.2018.01614

Dey, D., Nath, D., & Jamatia, P. B. (2015). Effect of raised bed planting method of maize under sandy loam soil of West Tripura. *International Journal of Applied Research, 1*(7), 561–563.

Dhima, K. V., Vasilakoglou, I. B., Keco, R. X., Dima, A. K., Paschalidis, K. A., & Gatsis, T. D. (2014). Forage yield and competition indices of faba bean intercropped with oat. *Grass and Forage Science, 69*, 376–383.

Dolferus, R., Ji, X., & Richards, R. A. (2011). Abiotic stress and control of grain number in cereals. *Plant Science, 181*, 331–341.

Dvořák, P., Capouchová, I., Král, M., Konvalina, P., Janovská, D., & Satranský, M. (2022). Grain yield and quality of wheat in wheat-legumes intercropping under organic and conventional growing systems. *Plant, Soil and Environment, 68*, 553–559.

Ehdaie, B., Whitkus, R. W., & Waines, J. G. (2003). Root biomass, water-use efficiency, and performance of wheat–rye translocations of chromosomes 1 and 2 in spring bread wheat 'Pavon.' *Crop Science, 43*, 710.

Ehsan, A. M., Ahmad, Z. K. S., & Bashir, K. (2002). Response of different wheat cultivars to potash application in two soil series of Pakistan. *Asian Journal of Plant Science, 5*, 535–537.

EL Sabagh, A., Islam, M. S., Skalicky, M., Raza, M. A., Singh, K., Hossain, M. A., & Arshad, A. (2021). Salinity stress in wheat (Triticum aestivum L.) in the changing climate: adaptation and management strategies. *Frontiers in Agronomy, 43.*

Elbashir, A. A. E., Gorafi, Y. S. A., Tahir, I. S. A., Kim, J. S., & Tsujimoto, H. (2017). Wheat multiple synthetic derivatives: A new source for heat stress tolerance adaptive traits. *Breeding Science, 67*(3), 248–256. https://doi.org/10.1270/jsbbs.16204

El-Hendawy, S. E., Hu, Y., & Schmidhalter, U. (2007). Assessing the suitability of various physiological traits to screen wheat genotypes for salt tolerance. *Journal of Integrative Plant Biology, 49*, 1352–1360.

EL-Shal, R. (2016). Effect of urea and potassium sulfate fertilizers combined with boron on soil fertility and sugar beet productivity in salt affected soil. *Egyptian Journal of Soil Science, 56*(4), 665–681.

Erdal, I., & Ekinci, K. (2020). Effects of composts and vermicomposts obtained from forced aerated and mechanically turned composting method on growth, mineral nutrition and nutrient uptake of wheat. *Journal of Plant Nutrition, 43*(9), 1343–1355.

Evans, L. T. (1993). *Crop evolution, adaptation and yield*. Cambridge University Press.

Evans, K. M., & Riedell, W. E. (2006). Response of spring wheat cultivars to nutrient solutions containing additional potassium chloride. *Journal of Plant Nutrition, 29*, 467–504.

Fan, F., Zhang, F., Song, Y., Sun, J., Bao, X., Guo, T., & Li, L. (2006). Nitrogen fixation of faba bean (Vicia faba L.) interacting with a non-legume in two contrasting intercropping systems. *Plant and Soil, 283*(1), 275–286.

Fan, Y., Wang, C., & Nan, Z. (2018). Determining water use efficiency of wheat and cotton: A meta-regression analysis. *Agricultural Water Management, 199*, 48–60.

Fan, Z., Chai, Q., Huang, G., Yu, A., Huang, P., Yang, C., Tao, Z., & Liu, H. (2013). Yield and water consumption characteristics of wheat/maize intercropping with reduced tillage in an Oasis region. *European Journal of Agronomy, 45*, 52–58.

FAO. (2012). Developing a climate-smart agriculture strategy at the country level: Lessons from recent experience. Background Paper for the Second Global Conference on Agriculture, Food Security and Climate Change; Food and Agriculture Organization of the United Nations.

FAO. (2014). The state of food insecurity in the world 2014. Strengthening the enabling environment for food security and nutrition. FAO.

Farkas, Z., Varga-László, E., Anda, A., Veisz, O., & Varga, B. (2020). Effects of waterlogging, drought and their combination on yield and water-use efficiency of five Hungarian winter wheat varieties. *Water, 12*, 1318.

Farooq, M., Wahid, A., Kobayashi, N., Fujita, D., & Basra, S. (2009). Plant drought stress: Effects, mechanisms and management. *Agronomy for Sustainable Development, 29*, 185–212.

Feldman, M. (2001). Origin of cultivated wheat. In A. P. Bonjean & W. J. Angus (Eds.), *The world wheat book: A history of wheat breeding* (pp. 3–56). Intercept Ltd.

Feng, Z., Liu, A., Yi, J., Li, R., Wang, X., & Zou, Q. (2014). Effects of amount of applying potassium fertilizer on yield, fiber quality and potassium fertilizer application efficiency of three hybrid cotton cultivars. *Journal of Hunan Agricultural University, 39*(4), 343–347.

Ferrie, A. M. R., & Caswell, K. L. (2011). Isolated microspore culture techniques and recent progress for haploid and doubled haploid plant production. *Plant Cell Tissue and Organ Culture, 104*, 301–309.

Figueroa, M., Hammond-Kosack, K. E., & Solomon, P. S. (2017). A review of plant diseases—A field perspective. *Molecular Plant Pathology, 19*, 1523–1536.

Figueroa-Bustos, V., Palta, J. A., Chen, Y., & Siddique, K. H. M. (2018). Characterization of root and shoot traits in wheat cultivars with putative differences in root system size. *Agronomy, 8*(7), 109. https://doi.org/10.3390/agronomy8070109

Fischer, R. A., Sayre, K., & Ortiz Monasterio, I. (2005). The effect of raised bed planting on irrigated wheat yield as influenced by variety and row spacing. In *Australian centre for international agriculture research proceeding of a workshop held in Griffith, NSW, Australia, 1–3* (pp. 121).

Ford, K. E., Gregory, P. J., Gooding, M. J., & Pepler, S. (2006). Genotype and fungicide effects on late-season root growth of winter wheat. *Plant and Soil, 284*, 33–34.

Foulkes, M. J., Snape, J. W., Shearman, V. J., Reynolds, M. P., Gaju, O., & Sylvester-Bradley, R. (2007). Genetic progress in yield potential in wheat: Recent advances and future prospects. *Journal of Agricultural Science, 145*, 17.

Franzen, D. W., Hopkins, D. H., Sweeney, M. D., Ulmer, M. K., & Halvorson, A. D. (2002). Evaluation of soil survey scale for zone development of site-specific nitrogen management. *Agronomy Journal, 94*, 381.

Fustec, J., Lesuffleur, F., Mahieu, S., & Cliquet, J. B. (2010). Nitrogen rhizodeposition of legumes. A review. *Agronomy for Sustainable Development, 30*, 57–66. https://doi.org/10.1051/agro/2009003

Galloway, J. N., Winiwarter, W., Leip, A., Leach, A. M., Bleeker, A., & Erisman, J. W. (2014). Nitrogen footprints: Past, present and future. *Environmental Research Letters, 9*, 115003. https://doi.org/10.1088/1748-9326/9/11/115003

Gan, Y. T., Liang, C., Hamel, C., Cutforth, H., & Wang, H. (2011). Strategies for reducing the carbon footprint of field crops for semiarid areas—A review. *Agronomy for Sustainable Development, 31*, 643–656. https://doi.org/10.1007/s13593-011-0011-7

George, T. S., Gregory, P. J., Hocking, P., & Richardson, A. E. (2008). Variation in root-associated phosphatase activities in wheat contributes to the utilization of organic P substrates in vitro, but does not explain differences in the P-nutrition of plants when grown in soils. *Environmental and Experimental Botany, 64*, 239–249. https://doi.org/10.1016/j.envexpbot.2008.05.002

Gill, B. S., Appels, R., Botha-Oberholster, A. M., Buell, C. R., Bennetzen, J. L., Chalhoub, B., Chumley, F., Dvovak, J., Iwanaga, M., Keller, B., Li, W., McCombie, W. R., Ogihara, Y., Quetier, F., & Sasaki, T. (2004). A workshop report on wheat genome sequencing: International genome research on wheat consortium. *Genetics, 168*, 1087–1096.

Goswami, L., Nath, A., Sutradhar, S., Bhattacharya, S. S., Kalamdhad, A., Vellingiri, K., et al. (2017). Application of drum compost and vermicompost to improve soil health, growth, and yield parameters for tomato and cabbage plants. *Journal of Environmental Management, 200*, 243–252.

Grant, C., Bittman, S., Montreal, M., Plenchette, C., & Morel, C. (2005). Soil and fertilizer phosphorus: Effects on plant P supply and mycorrhizal development. *Canadian Journal of Plant Science, 85*, 3–14.

Grant, C. A., Flaten, D. N., Tomasiewicz, D. J., & Sheppard, S. C. (2001). The importance of early season phosphorus nutrition. *Canadian Journal of Plant Science, 81*(2), 211–224.

Gregory, P. J., & Brown, S. C. (1989). Root growth, water use and yield of crops in dry environments: What characteristics are desirable? *Aspects of Applied Biology, 22*, 234–243.

Groppa, M., & Benavides, M. (2008). Polyamines and abiotic stress: Recent advances. *Amino Acids, 34*, 35–45.

Gu, X., Liu, Y., Li, N., Liu, Y., Zhao, D., Wei, B., & Wen, X. (2021). Effects of the foliar application of potassium fertilizer on the grain protein and dough quality of wheat. *Agronomy, 11*, 1749. https://doi.org/10.3390/agronomy11091749

Gupta, N. K., Gupta, S., & Kumar, A. (2001). Effect of water stress on physiological attributes and their relationship with growth and yield in wheat cultivars at different growth stages. *J. Agron, 86*, 1437–1439.

Gupta, A., Rico-Medina, A., & Cano-Delgado, A. I. (2020). The physiology of plant responses to drought. *Science, 368*, 266–269.

Gupta, V., Rai, P. K., & Risam, K. S. (2012). Integrated crop-livestock farming systems: A strategy for resource conservation and environmental sustainability. *India Research Journal of Extension Education, S2*, 49–54.

Habash, D. Z., Baudo, M., Hindle, M., Powers, S. J., Defoin-Platel, M., Mitchell, R., et al. (2014). Systems responses to progressive water stress in durum wheat. *PloS ONE, 9*, e108431. https://doi.org/10.1371/journal.pone.0108431

Haji, A., Bu, M. A. A., Ahmad, R., Iqbal, J., Maqbool, M. M., Ali, A., Ishaque, M., & Hussain, S. (2012). Nutritional and physiological significance of potassium application in maize hybrid crop production. *Pakistan Journal of Nutrition, 11*(2), 187–202.

Hameed, A., Goher, M., & Iqbal, N. (2012). Heat stress-induced cell death, changes in antioxidants, lipid peroxidation, and protease activity in wheat leaves. *Journal of Plant Growth Regulation, 31*, 283–291. https://doi.org/10.1007/s00344-011-9238-4

Hammad, H. M., Khaliq, A., Ahmad, A., Aslam, M., Malik, A. H., Farhad, W., & Laghari, K. Q. (2017) Influence of different organic manures on wheat productivity. *International Journal of Agriculture and Biology, 13*(1), 137–140.

Hasanuzzaman, M., Hossain, M. A., & Fujita, M. (2011). Selenium-induced up-regulation of the antioxidant defense and methylglyoxal detoxification system reduces salinity-induced damage in rapeseed seedlings. *Biological Trace Element Research, 143*, 1704–1721. https://doi.org/10.1007/s12011-011-8958-4

Hassan, I., Hussain, Z., & Akbar, G. (2005). Effect of permanent raised beds on water productivity for irrigated maize–Wheat cropping system. In *Evaluation and performance of permanent raised bed cropping systems in Asia, Australia and Mexico. ACIAR Proceeding* (Vol. 121). https://ageconsearch.umn.edu/record/135397/files/PR121.pdf#page=64

Hauggaard-Nielsen, H., Ambus, P., & Jensen, E. S. (2001). Interspecific competition, N use and interference with weeds in pea–barley intercropping. *Field Crops Research, 70*(2), 101–109.

Havlin, J. L., Tisdale, S. L., Nelson, W. L., & Beaton, J. D. (2016). *Soil fertility and fertilizers. An introduction to nutrient management* (7th Ed). Prentice Hall of India.

Hawkesford, M. J., Araus, J. L., Park, R., Calderini, D., Miralles, D., Shen, T., Zhang, J., & Parry, M. A. J. (2013). Prospects of doubling global wheat yields. *Food Energy Security, 2*, 34–48.

Hedge, D., & Murthy, I. (2005). Management of secondary nutrients. *Indian Journal of Fertilisers, 9*, 90–100.

Huang, B., Rachmilevitch, S., & Xu, J. (2012). Root carbon and protein metabolism associated with heat tolerance. *Journal of Experimental Botany, 63*(9), 3455–3465.

Hussain, I., Ali, A., Ahmed, A., Nasrullah, H., Iqbal, S., Aulakh, A. M., & Ahmed, G. (2018). Impact of ridge-furrow planting in Pakistan: Empirical evidence from the farmers field. *International Journal of Agronomy*. https://doi.org/10.1155/2018/3798037

Hussain, N., Khan, M. B., & Ahmad, R. (2008). Influence of phosphorus application and sowing time on performance of wheat in calcareous soils. *International Journal of Agriculture and Biology, 10*, 399–404.

Hussain, N., Ghaffar, A., Zafar, Z. U., Javed, M., Shah, K. S., Noreen, S., Manzoor, H., Iqbal, M., Hassan, I. F. Z., Bano, H., Gul, H. S., Aamir, M., Khalid, A., Sohail, Y., Ashraf, M., & Athar, H. U. R. (2021). Identification of novel source of salt tolerance in local bread wheat germplasm using morpho-physiological and biochemical attributes. *Scientific Reports, 11*, 10854.

Hussain, S., Shah, M. A. A., Khan, A. M., Ahmad, F., & Hussain, M. (2020). Potassium enhanced grain zinc accumulation in wheat grown on a calcareous saline-sodic soil. *Pakistan Journal of Botany, 52*(1), 69–74.

Ibrahim, M., Ullah, H., Ahmad, B., & Malik, M. F. A. (2012). Effect of incremental dose of phosphorous and sulphur upon yield and protein content of wheat. *Biological Diversity and Conservation, 5*(3), 76–81.

Idnani, L. K., & Kumar, A. (2013). Performance of wheat (Triticum aestivum) under different irrigation schedules and sowing methods. *The Indian Journal of Agricultural Sciences, 83*(1), 37–40.

Igbadun, H. E., Mahoo, H. F., Andrew, K. P. R., Baanda, T., & Salim, A. (2006). Crop water productivity of an irrigated maize crop in Mkoji sub-catchment of the Great Ruaha River Basin, Tanzania. *Agricultural Water Management, 85*(1–2), 141–150.

Izanloo, A., Condon, A. G., Langridge, P., Tester, M., & Schnurbusch, T. (2008). Different mechanisms of adaptation to cyclic water stress in two South Australian bread wheat cultivars. *Journal of Experimental Botany, 59*, 3327–3346. https://doi.org/10.1093/jxb/ern199

Jan, T., Jan, M. T., Arif, M., Akbar, H., & Ali, S. (2007). Response of wheat to source, type and time of nitrogen application. *Sarhad Journal of Agriculture, 23*(4), 871.

Jensen, E. S., Carlsson, G., & Hauggaard-Nielsen, H. (2020). Intercropping of grain legumes and cereals improves the use of soil N resources and reduces the requirement for synthetic fertilizer N: A global-scale analysis. *Agronomy for Sustainable Development, 40*, 5.

Jighly, A., Alagu, M., Makdis, F., Singh, M., Singh, S., Emebiri, L. C., & Ogbonnaya, F. C. (2016). Genomic regions conferring resistance to multiple fungal pathogens in synthetic hexaploid wheat. *Molecular Breeding, 36*, 127.

Joshi, R., Vig, A. P., & Singh, J. (2013). Vermicompost as soil supplement to enhance growth, yield and quality of Triticum aestivum L.: a field study. *International Journal of Recycling of Organic Waste in Agriculture, 2*(1), 1–7.

Kabato, W., Ergudo, T., Mutum, et al. (2022). Response of wheat to combined application of nitrogen and phosphorus along with compost. *Journal of Crop Science and Biotechnology, 25*, 557–564. https://doi.org/10.1007/s12892-022-00151-7

Katare, S., Singh, B., Patil, S. D., Tiwari, R., Jasrotia, P., Saharan, M. S., & Sharma, I. (2015). Evaluation of new insecticides for management of foliar aphid complex in wheat. *Indian Journal of Entomology, 79*(2), 185–190.

Kazi, A. V., Rasheed, A., & Mujeeb-Kazi, I. A. (2013). Biotic stress and crop improvement: A wheat focus around novel strategies. In K. R. Hakeem, et al. (Eds.), *Crop improvement*. https://doi.org/10.1007/978-1-4614-7028-1_7. © Springer Science+Business Media, LLC.

Kirigwi, F., van Ginkel, M., Trethowan, R., Sears, R. G., Rajaram, S., & Paulsen, G. M. (2004). Evaluation of selection strategies for wheat adaptation across water regimes. *Euphytica, 135*, 361–371.

Kirkegaard, J. A., Christen, O., Krupinsky, J., & Layzell, D. (2008). Break crop benefits in temperate wheat production. *Field Crops Research, 107*, 185–195.

Kizilkaya, R. F., Turkay, S. H., Turkmen, C., & Durmus, M. (2012). Vermicompost effects on wheat yield and nutrient contents in soil and plant. *Archives of Agronomy and Soil Science, 58*(supl), S175–S179. https://doi.org/10.1080/03650340.2012.696777

Koca, H., Bor, M., Ozdemir, F., & Turkan, I. (2007). The effect of salt stress on lipid peroxidation, antioxidative enzymes and proline content of sesame cultivars. *Environmental and Experimental Botany, 60*, 344–351.

Kröbel, R., Lemke, R., Campbell, C. A., Zentner, R., McConkey, B., Steppuhn, H., De Jong, R., & Wang, H. (2014). Water use efficiency of spring wheat in the semi-arid Canadian prairies: Effect of legume green manure, type of spring wheat, and cropping frequency. *Canadian Journal of Soil Science, 94*, 223–235.

Kukal, M. S., & Irmak, S. (2018). Climate-driven crop yield and yield variability and climate change impacts on the US Great Plains agricultural production. *Scientific Reports, 8*(1), 3450. https://doi.org/10.1038/s41598-018-21848-2. PMID: 29472598.

Kumari, R., Kaushal, A., & Singh, K. G. (2014). Water use efficiency of drip fertigated sweet pepper under the influence of different kinds and levels of fertilizers. *Indian Journal of Science and Technology, 7*(10), 1538–1543.

Kutman, U. B., Yildiz, B., & Cakmak, I. (2011). Effect of nitrogen on uptake, remobilization and partitioning of zinc and iron throughout the development of durum wheat. *Plant and Soil, 342*(1), 149–164.

Ladha, J. K., Pathak, H., Krupnik, T. J., Six, J., & van Kessel, C. (2005). Efficiency of fertilizer nitrogen in cereal production: Retrospects and prospects. *Advances in Agronomy., 87*, 85–156.

Latati, M., Bargaz, A., Belarbi, B., Lazali, M., Benlahrech, S., Tellah, S., Kaci, G., Drevon, J. J., & Ounane, S. M. (2016). The intercropping common bean with maize improves the rhizobial efficiency, resource use and grain yield under low phosphorus availability. *European Journal of Agronomy, 72*, 80–90. https://doi.org/10.1016/j.eja.2015.09.015

Lawson, Y. D., Dzomeku, I., & Drisah, Y. (2007). Time of planting Mucuna and Canavalia in an intercrop system with maize. *Journal of Agronomy, 6*, 534. https://doi.org/10.3923/ja.2007.534.540

Leng, G., & Hall, J. (2019). Crop yield sensitivity of global major agricultural countries to droughts and the projected changes in the future. *Science of the Total Environment, 654*, 811–821.

Li, Q., Bian, C., Liu, X., Ma, C., & Liu, Q. (2015). Winter wheat grain yield and water use efficiency in wide-precision planting pattern under deficit irrigation in North China Plain. *Agricultural Water Management, 153*, 71–76.

Liang, X., & Yu, Z. (2004). Effect of potassium application stage on photosynthetic characteristics of winter wheat flag leaves and on starch accumulation in wheat grains. *Journal of Applied Ecology, 15*(8), 1349–1352.

Lilley, J. M., & Kirkegaard, J. A. (2011). Benefits of increased soil exploration by wheat roots. *Field Crops Research, 122*, 118–130.

Limon-Ortega, A., Sayre, K. D., Drijber, R. A., & Francis, C. A. (2002). Soil attributes in a furrow-irrigated bed planting system in northwest Mexico. *Soil and Tillage Research, 63*, 123–132.

Liu, C. A., Li, F. M., Zhou, L. M., Zhang, R. H., Lin, S. L., Wang, L. J., et al. (2013). Effect of organic manure and fertilizer on soil water and crop yields in newly-built terraces with loess soils in a semi-arid environment. *Agricultural Water Management, 117*, 123–132. https://doi.org/10.1016/j.agwat.2012.11.002

Liu, H. W., Du, X. F., Li, Y. B., Han, X., Li, B., Zhang, X. K., Li, Q., Liang, W. J. (2022). Organic substitutions improve soil quality and maize yield through increasing soil microbial diversity. *Journal of Cleaner Production, 347*, Article 131323.

Liu, J., Zhang, W., Long, S., & Zhao, C. (2021). Maintenance of cell wall integrity under high salinity. *International Journal of Molecular Sciences, 22*, 3260.

Lodhi, A. S., Kaushal, A., & Singh, K. G. (2014). Impact of irrigation regimes on growth, yield and water use efficiency of sweet pepper. *Indian Journal of Science and Technology., 7*(6), 790–794.

Lollato, R. P., Figueiredo, B. M., Dhillon, J. S., Arnall, D. B., & Raun, W. R. (2019). Wheat grain yield and grain-nitrogen relationships as affected by N, P, and K fertilization: A synthesis of long-term experiments. *Field Crops Research, 236*, 42–57.

Lopes, M. S., & Reynolds, M. P. (2010). Partitioning of assimilates to deeper roots is associated with cooler canopies and increased yield under drought in wheat. *Functional Plant Biology, 37*, 147–156.

Lu, Q., Jia, D., Zhang, Y., Dai, X., He, M., et al. (2014). Split application of potassium improves yield and end-use quality of winter wheat. *Agronomy Journal, 106*(4), 1411–1419.

Lynch, J. P., & Brown, K. M. (2001). Topsoil foraging-an architectural adaptation of plants to low phosphorus availability. *Plant and Soil, 237*, 225–237. https://doi.org/10.1023/A:1013324727040

Majeed, A., Muhmood, A., Niaz, A. N., Javid, S., Ahmed, Z. A., Shah, S. S. H., & Shah, A. H. (2015). Bed planting of wheat (Triticum aestivum L.) improves nitrogen use efficiency and grain yield compared to flat planting. *The Crop Journal, 3*, 118–124.

Malve, S. H., Rao, P. V., & Dhake, A. (2017). Yield and nutrient uptake of drip irrigated wheat under varying irrigation and nitrogen levels. *Multilogic in Science, VII*(XXIV), 21–25.

Mao, L. L., Zhang, L. Z., Li, W. Q., van der Werf, W., Sun, J. H., Spiertz, H., & Li, L. (2012). Yield advantage and water saving in maize/pea intercrop. *Field Crops Research, 138*(11–20), 48.

Marček, T., Áaron, K., Végh, B., Janda, T., & Darko, E. (2019). Metabolic response to drought in six winter wheat genotypes. *PloS ONE, 14*, e0212411.

Martins, M. Q., Rodrigues, W. P., Fortunato, A. S., Leitao, A. E., Rodrigues, A. P., Pais, I. P., Martins, L. D., Silva, M. J., Reboredo, F. H., Partelli, F. L., & Campostrini, E. (2016). Protective response mechanisms to heat stress in interaction with high [CO_2] conditions in Coffee spp. *Frontiers in Plant Science, 7*, 947.

Martre, P., Wallach, D., Asseng, S., Ewert, F., Jones, J. W., Rötter, R. P., et al. (2015). Multi model ensembles of wheat growth: Many models are better than one. *Global Change Biology., 21*(2), 911–925.

Masaka, J. (2006). The effect of n fertilizer placement and timing on the aboveground biometric characteristics of spring wheat (Triticum aestivum L. cv. Spectrum) on Leached Chernozem. *International Journal of Agricultural Research, 1*, 68–75.

Mehdi, S. M., Ranjha, A. M., Sarfraz, M., & Hassan, G. (2001). Response of wheat to potassium application in six soil series of Pakistan. *Journal of Biological Sciences, 6*, 429–431.

Mehriya, M. L., Geat, N., Sarita, Singh, H., Mattar, M. A., & Elansary, H. O. (2020). Response of drip irrigation and fertigation on cumin yield, quality, and water-use efficiency grown under arid climatic conditions. *Agronomy, 10*(1711), 1–16.

Mei, X. R., Zhong, X. L., Vadez, V., & Liu, X. Y. (2013). Improving water use efficiency of wheat crop varieties in the North China Plain: Review and analysis. *Journal of Integrative Agriculture, 12*, 1243–1250.

Menshawy, A. M. M. (2007). Evaluation of some early bread wheat genotypes under different sowing dates: 1 Earliness characters. *Egyptian Journal of Plant Breeding, 11*(1), 25–40.

Mesfin, S., Gebresamuel, G., Haile, M., & Zenebe, A. (2023). Potentials of legumes rotation on yield and nitrogen uptake of subsequent wheat crop in northern Ethiopia. *Heliyon., 9*(6), e16126. https://doi.org/10.1016/j.heliyon.2023.e16126.PMID:37292361;PMCID:PMC10245104

Miah, M. A. M., Hossain, S., Duxbury, J. M., & Lauren, J. G. (2015). Adoption of raised bed technology in some selected locations of Rajshahi District of Bangladesh. *Bangladesh Journal of Agricultural Research, 40*(4), 551–566. https://doi.org/10.3329/bjar.v40i4.26930

Mickelbart, M., Hasegawa, P., & Bailey-Serres, J. (2015). Genetic mechanisms of abiotic stress tolerance that translate to crop yield stability. *Nature Reviews Genetics, 16*, 237–251. https://doi.org/10.1038/nrg3901

Miller, O., Helman, D., Svoray, T., Morin, E., & Bonfil, D. J. (2019). Explicit wheat production model adjusted for semi-arid environments. *Field Crops Research, 231*, 93–104.

Mirzaei, A., Naseri, R., & Soleimani, R. (2011). Response of different growth stages of wheat to moisture tension in a semiarid land. *World Applied Sciences Journal, 12*(1), 83–89.

Mohamed, M. F., Thalooth, A. T., Elewa, T. A., et al. (2019). Yield and nutrient status of wheat plants (Triticum aestivum) as affected by sludge, compost, and biofertilizers under newly reclaimed soil. *Bulletin of the National Research Centre, 43*, 31. https://doi.org/10.1186/s42269-019-0069-y

Molden, D., Oweis, T. Y., Steduto, P., Kijne, J. W., Hanjra, M. A., Bindraban, P. S., Bouman, B. A. M., et al. (2007). Pathways for increasing agricultural water productivity. In *Water for food water for life: A comprehensive assessment of water management in agriculture*. Taylor and Francis.

Mondini, L., & Pagnotta, M. A. (2015). Sustainable agriculture reviews. E. Lichtfouse, & A. Goyal (Eds.). Springer International Publishing.

Morales, F., Ancín, M., Fakhet, D., González-Torralba, J., Gámez, A. L., Seminario, A., Soba, D., Ben Mariem, S., Garriga, M., & Aranjuelo, I. (2020). Photosynthetic metabolism under stressful growth conditions as a bases for crop breeding and yield improvement. *Plants, 9*, 88. https://doi.org/10.3390/plants9010088

Nassar, R., Kamel, H. A., Ghoniem, A. E., Alarcón, J. J., Sekara, A., Ulrichs, C., et al. (2020). Physiological and anatomical mechanisms in wheat to cope with salt stress induced by seawater. *Plants, 9*, 237. https://doi.org/10.3390/plants9020237

Negrão, S., Schmöckel, S. M., & Tester, M. (2017). Evaluating physiological responses of plants to salinity stress. *Annals of Botany, 119*, 1–11.

Nguyen, H. T., Nguyen, A. T., Lee, B. W., & Schoenau, J. (2002). Effects of long-term fertilization for cassava production on soil nutrient availability as measured by ion exchange membrane probe and by corn and canola nutrient uptake. *Korean Journal of Crop Science, 47*, 108–115.

Nieder, R., & Benbi, D. K. (2008). *Carbon and nitrogen in the terrestrial environment*. Springer.

Nisar, A. (2000). Integrated plant nutrition management in Pakistan: Status and opportunities. In *Proceeding of symposium on integrated plant nutrition management, NFDC, Islamabad, pp. 18–39.*

Niu, J., Zhang, W., Ru, S., Chen, X., Xiao, K., Zhang, X., et al. (2013). Effects of potassium fertilization on winter wheat under different production practices in the North China Plain. *Field Crops Research, 1*(140), 69–76.

Nuruzzaman, M., Lambers, H., Bolland, M. D. A., & Veneklaas, E. J. (2005). Phosphorus benefits of different legume crops to subsequent wheat grown in different soils of Western Australia. *Plant and Soil, 271*, 175–187. https://doi.org/10.1007/s11104-004-2386-6

Ogbonnaya, F. C., Abdalla, O., Mujeeb-Kazi, A., Kazi, A. G., Xu, S. S., Gosman, N., Lagudah, E. S., Bonnett, D., Sorrells, M. E., & Tsujimoto, H. (2013). Synthetic hexaploids: Harnessing species of the primary gene pool for wheat improvement. *Plant Breeding Reviews, 37*, 35–122.

Ooro, P. A., Malinga, J. N., Tanner, D. G., & Payne, T. S. (2011). Implication of rate and time of nitrogen application on wheat (Triticum aestivum L.) yield and quality in Kenya. *Journal of Animal and Plant Sciences, 9*(2), 1141–1146.

Osakabe, Y., Osakabe, K., Shinozaki, K., & Tran, L.-S.P. (2014). Response of plants to water stress. *Frontiers in Plant Science, 5*, 86.

Palta, J. A., & Yang, J. (2014). Crop root system behaviour and yield. *Field Crops Research, 165*, 1–4.

Pandey, V., & Shukla, A. (2015). Acclimation and tolerance strategies of rice under drought stress. *Rice Science, 22*, 147–161. https://doi.org/10.1016/j.rsci.2015.04.001

Pankou, C., Lithourgidis, A., & Dordas, C. (2021). Effect of irrigation on intercropping systems of wheat (Triticum aestivum L.) with Pea (Pisum sativum L.). *Agronomy, 11*(2), 283. https://doi.org/10.3390/agronomy11020283

Pelzer, E., Bazot, M., Makowski, D., Corre-Hellou, G., Naudin, C., Al, R. M., Baranger, E., Bedoussac, L., Biarnes, V., Boucheny, P., Carrouée, B., Dorvillez, D., Foissy, D., Gaillard, B., Guichard, L., Mansard, M.-C., Omon, B., Prieur, L., Yvergniaux, M., … Jeuffroy, M. H. (2012). Pea-wheat intercrops in low input conditions combine high economic performances and low environmental impacts. *European Journal of Agronomy, 40*, 39–53.

Péret, B., Clément, M., Nussaume, L., & Desnos, T. (2011). Root developmental adaptation to phosphate starvation: Better safe than sorry. *Trends in Plant Science, 16*, 442–450. https://doi.org/10.1016/j.tplants.2011.05.006

Pettigrew, W. T. (2008). Potassium influences on yield and quality production for maize, wheat, soybean and cotton. *Physiologia Plantarum, 133*, 670–681.

Pinto, R. S., Reynolds, M. P., Mathews, K. L., McIntyre, C. L., Olivares-Villegas, J. J., & Chapman, S. C. (2010). Heat and drought adaptive QTL in a wheat population designed to minimize confounding agronomic effects. *TAG. Theoretical and Applied Genetics, 121*, 1001–1021.

Poiroux-Gonord, F., Santini, J., Fanciullino, A., Lopez Lauri, F., Giannettini, J., Sallanon, H., et al. (2013). Metabolism in orange fruits is driven by photooxidative stress in leaves. *Physiologia Plantarum, 149*, 175187.

Preissel, S., Reckling, M., Schläfke, N., & Zander, P. (2015). Magnitude and farm-economic value of grain legume pre-crop benefits in Europe: A review. *Field Crops Research, 175*, 64. https://doi.org/10.1016/j.fcr.2015.01.012

Rao, K. V. R., Bajpai, A., Gangwar, S., Chourasia, L., & Soni, K. (2016). Maximising water productivity of wheat crop by adopting drip irrigation. *Research on Crops, 17*, 163–168.

Ray, D. K., Ramankutty, N., Mueller, N. D., West, P. C., & Foley, J. A. (2012). Recent patterns of crop yield growth and stagnation. *Nature Communications, 3*, 1293.

Reganold, J. P., & Wachter, J. M. (2016). Organic agriculture in the twenty-first century. *Nature Plants, 2*(2), 15221. https://doi.org/10.1038/nplants.2015.221

Reynolds, M., Bonnett, D., Chapman, S. C., Furbank, R. T., Manes, Y., Mather, D. E., & Parry, M. A. J. (2011). Raising yield potential of wheat. I. Overview of a consortium approach and breeding strategies. *Journal of Experimental Botany, 62*, 439–452.

Reynolds, M., Foulkes, M. J., Slafer, G. A., Berry, P., Parry, M. A. J., Snape, J. W., & Angus, W. J. (2009). Raising yield potential in wheat. *Journal of Experimental Botany, 60*, 1899–1918.

Richardson, A. E., Hocking, P. J., Simpson, R. J., & George, T. S. (2009). Plant mechanisms to optimize access to soil phosphorus. *Crop & Pasture Science, 60*, 124–143. https://doi.org/10.1071/CP07125

Richardson, A. E., Lynch, J. P., Ryan, P. R., Delhaize, E., Smith, F. A., Smith, S. E., et al. (2011). Plant and microbial strategies to improve the phosphorus efficiency of agriculture. *Plant and Soil, 349*, 121–156. https://doi.org/10.1007/s11104-011-0950-4

Roberts, T. L. (2008). Improving nutrient use efficiency. *Turkish Journal of Agriculture and Forestry, 32*, 177–182.

Robinson, D. (2001). Root proliferation, nitrate inflow and their carbon costs during nitrogen capture by competing plants in patchy soil. *Plant and Soil, 232*, 41–50.

Rollins, J. A., Habte, E., Templer, S. E., Colby, T., Schmidt, J., & von Korff, M. (2013). Leaf proteome alterations in the context of physiological and morphological responses to drought and heat stress in barley (Hordeum vulgare L.). *Journal of Experimental Botany, 64*(11), 3201–3212.

Romheld, V., & Kirkby, E. A. (2010). Research on potassium in agriculture: Needs and prospects. *Plant and Soil, 335*, 155–180.

Sallam, A., Alqudah, A. M., Dawood, M. F. A., Baenziger, P. S., & Börner, A. (2019). Drought stress tolerance in wheat and barley: Advances in physiology, breeding and genetics research. *International Journal of Molecular Sciences, 20*(13), 3137. https://doi.org/10.3390/ijms20133137.PMID:31252573;PMCID:PMC6651786

Sardans, J., & Peñuelas, J. (2015). Potassium: A neglected nutrient in global change. *Global Ecology and Biogeography, 24*, 261–275.

Sarkar, S., Aminul Islam, A. K. M., Barma, N. C. D., & Ahmed, J. U. (2021). Tolerance mechanisms for breeding wheat against heat stress: A review. *South African Journal of Botany, 138*, 1–16. https://doi.org/10.1016/j.sajb.2021.01.003

Seymour, M., Kirkegaard, J., Peoples, M., White, P., & French, R. (2012). Break-crop benefits to wheat in Western Australia–insights from over three decades of research. *Crop & Pasture Science, 63*, 1–16. https://doi.org/10.1071/CP11320

Shabnam, R., & Iqbal, M. T. (2016). Phosphorus use efficiency by wheat plants that grown in an acidic soil. *Brazilian Journal of Science and Technology, 3*, 18. https://doi.org/10.1186/s40552-016-0030-7

Shah, J. A., Depar, N., Memon, M. Y., Aslam, M., & Sial, N. A. (2013). Integration of organic and mineral nutrient sources enhances wheat production. *Pakistan Journal of Agriculture, Agricultural Engineering and Veterinary Sciences, 29*, 106–113.

Shahzad, Z., & Amtmann, A. (2017). Food for thought: How nutrients regulate root system architecture. *Current Opinion in Plant Biology., 39*, 80–87.

Shanahan, J. F., Kitchen, N. R., Raun, W. R., & Schepers, J. S. (2008). Responsive in-season nitrogen management for cereals. *Computers and Electronics in Agriculture, 61*(1), 51–62.

Shi, R., Zhang, Y., Chen, X., Sun, Q., Zhang, F., Romheld, V., & Zou, C. (2010). Influence of long-term nitrogen fertilization on micronutrient density in grain of winter wheat (Triticum aestivum L.). *Journal of Cereal Science, 51*(1), 165–170.

Sing, V. K., Dwivedi, B. S., Shukla, A. K., & Mishra, R. P. (2010). Permanent raised bed planting of the pigeon pea-wheat system on a Typic Ustoochrept: Effects on soil fertility, yield and water and nutrient use efficiencies. *Field Crops Research, 116*, 127–139.

Singh, B., Aulakh, C., & Walia, S. (2019). Productivity and water use of organic wheat–chickpea intercropping system under limited moisture conditions in Northwest India. *Renewable Agriculture and Food Systems, 34*(2), 134–143. https://doi.org/10.1017/S1742170517000370

Smith, S., & De Smet, I. (2012). Root system architecture: Insights from Arabidopsis and cereal crops. *Philosophical Transactions of the Royal Society of London: Biological Sciences, 367*, 1441–1452.

Smith, S. E., & Read, D. J. (2010). *Mycorrhizal symbiosis*. Academic Press.

Sohail, Q., Inoue, T., Tanaka, H., Elsadig Eltayeb, A., Matsuoka, Y., & Tsujimoto, H. (2011). Applicability of Aegilops tauschii drought tolerance traits to breeding of hexaploid wheat. *Breeding Science, 61*, 347–357.

Stomph, T., Dordas, C., Baranger, A., de Rijk, J., Dong, B., Evers, J., Gu, C., Li, L., Simon, J., Jensen, E. S., Wang, Q., Wang, Y., Wang, Z., Xu, H., Zhang, C., Zhang, L., Zhang, W. P., Bedoussac, L., & van der Werf, W. (2020). Designing intercrops for high yield, yield stability and efficient use of resources: Are there principles? *Advances in Agronomy, 160*, 1–50.

Sunil, B., Saini, D., Bapatla, R. B., et al. (2019). Photorespiration is complemented by cyclic electron flow and the alternative oxidase pathway to optimize photosynthesis and protect against abiotic stress. *Photosynthesis Research, 139*, 67–79. https://doi.org/10.1007/s11120-018-0577

Suryavanshi, P., & Buttar, G. S. (2018). Effect of water saving micro-irrigation techniques on growth and yield of wheat in North West India. *Journal of Soil and Water Conservation, 17*(3), 250–258.

Suso, M. J., & Maalouf, F. (2010). Direct and correlated responses to upward and downward selection for outcrossing in Vicia faba. *Field Crops Research, 116*(1–2), 116. https://doi.org/10.1016/j.fcr.2009.12.001

Syers, J., Johnston, A., & Curtin, D. (2008). *Efficiency of soil and fertilizer phosphorus use*. FAO Fertilizer and Plant Nutrition Bulletin.

Taher, F. M., & AL-Naser, H. H. (2021). The effect of different levels of potassium on the productivity of genotypes of wheat Triticum Aestivum L. *Earth and Environmental Science, 923*(2021), 012061.

Talebi, R. (2011). Evaluation of chlorophyll content and canopy temperature as indicators for drought tolerance in durum wheat (Triticum durum Desf.). *Australian Journal of Basic and Applied Sciences, 5*(11), 1457–1462.

Talukder, A. S., McDonald, G. K., & Gill, G. S. (2014). Effect of short-term heat stress prior to flowering and early grain set on the grain yield of wheat. *Field Crops Research, 160*, 54–63.

Tan, M. X., Gou, F., Stomph, T. J., Wang, J., Yin, W., Zhang, L. Z., Chai, Q., & Van der Werf, W. (2020). Dynamic process-based modelling of crop growth and competitive water extraction in relay strip intercropping: Model development and application to wheat–maize intercropping. *Field Crops Research, 246*, 107613.

Tang, X., Bernard, L., Brauman, A., Daufresne, T., Deleporte, P., Desclaux, D., Souche, G., Placella, S. A., & Hinsinger, P. (2014). Increase in microbial biomass and phosphorus availability in the rhizosphere of intercropped cereal and legumes under field conditions. *Soil Biology & Biochemistry, 75*, 86–93. https://doi.org/10.1016/j.soilbio.2014.04.001

Tembo, B., Mulenga, R. M., Sichilima, S., Msiska, K. K., Mwale, M., & Chikoti, P. C. (2020). Detection and characterization of fungus (Magnaportheoryzae pathotype Triticum) causing wheat blast disease on rain-fed grown wheat (Triticum aestivum L.) in Zambia. *PlosOne, 15*(9), e0238724.

Thomas, H., & Ougham, H. (2014). The stay-green trait. *Journal of Experimental Botany, 65*(14), 3889–3900. https://doi.org/10.1093/jxb/eru037

Tilman, D., Fargione, J., Wolff, B., D'Antonio, C., Dobson, A., Howarth, R., Schindler, D., Schlesinger, W., Simberloff, D., & Swackhamer, D. (2001). Forecasting agriculturally driven global environmental change. *Science, 292*, 281–284.

Tuna Levent, A., Kaya, C., Dikilitas, M., & Higgs, D. (2008). The combined effects of gibberellic acid and salinity on some antioxidant enzyme activities, plant growth parameters and nutritional status in maize plants. *Environmental and Experimental Botany, 62*, 1–9.

UNEP. (2019). Frontiers 2018/19: Emerging issues of environmental concern. UNEP.

Vance, C., Ehde-Stone, C., & Allan, D. (2003). Phosphorus acquisition and use: Critical adaptations by plants for securing a nonrenewable resource. *New Phytologist, 157*, 423–447.

Vandermeer, J., Van Noordwijk, M., Anderson, J., & Perfecto, I. (1998). Global change and multi-species agroecosystems: Concepts and issues. *Agriculture, Ecosystems & Environment, 67*, 1–22.

Vanitha, K. (2008). *Drip fertigation and its physiological impact in aerobic rice (Oryzasativa L.)*. MSc. (Ag.) Thesis submitted to Tamil Nadu Agricultural University, Coimbatore, India, pp. 250.

Vereijken, P., & Kloen, H. (1994). Innovative research with iconological pilot farmers. In *Plant production on the threshold of a new century. Proceedings of the international conference at the occasion of the 7th anniversary of the Wageningen agricultural university, held between June 28 and July 1, 1993, Wageningen, The Netherlands.*

Verma, S., Nizam, S., & Verma, P. K. (2013). Biotic and abiotic stress signaling in plants. *Stress Signaling in Plants: Genomics and Proteomics Perspective, 1*, 25–49.

Vijayalakshmi, K., Fritz, A. K., Paulsen, G. M., Bai, G., Pandravada, S., & Gill, B. S. (2010). Modeling and mapping QTL for senescence-related traits in winter wheat under high temperature. *Molecular Breeding, 26*, 163–175.

Wang, J., Liu, W. Z., & Dang, T. H. (2011). Responses of soil water balance and precipitation storage efficiency to increased fertilizer application in winter wheat. *Plant and Soil, 347*(1–2), 41–51. https://doi.org/10.1007/s11104-011-0764-4

Wang, L., Wang, S., Chen, W., Li, H., & Deng, X. (2017). Physiological mechanisms contributing to increased water-use efficiency in winter wheat under organic fertilization. *PLoS ONE, 12*(6), e0180205. https://doi.org/10.1371/journal.pone.0180205.PMID:28662113; PMCID:PMC5491151

Wasson, A. P., Richards, R. A., Chatrath, R., Misra, S. C., Prasad, S. V., Rebetzke, G. J., Kirkegaard, J. A., Christopher, J., & Watt, M. (2012). Traits and selection strategies to improve root systems and water uptake in water-limited wheat crops. *Journal of Experimental Botany, 63*, 3485–3498.

Wellings, C. R. (2011). Global status of stripe rust: A review of historical and current threats. *Euphytica, 179*, 129–141.

Wick, A., Berti, M., Lawley, Y., & Liebig, M. (2017). Integration of Annual and Perennial Cover Crops for Improving Soil Health. In *Soil Health and Intensification of Agroecosytems* (pp. 127–150). Elsevier BV: Amsterdam, The Netherlands.

Xie X., He Z., Chen N., Tang Z., Wang Q., & Cai Y. (2019). The roles of environmental factors in regulation of oxidative stress in plant. *BioMed Research International, 9732325*. https://doi.org/10.1155/2019/9732325

Xiong, W., Reynolds, M. P., Crossa, J., et al. (2021). Increased ranking change in wheat breeding under climate change. *Nature Plants, 7*, 1207–1212. https://doi.org/10.1038/s41477-021-00988-w

Yang, X., Ni, K., Shi, Y., Yi, X., Zhang, Q., Fang, L., Ma, L., & Ruan, J. (2018). Effects of long-term nitrogen application on soil acidification and solution chemistry of a tea plantation in China. *Agriculture, Ecosystems & Environment, 252*, 74–82.

Yang, X., Zhang, C., Ma, X., Liu, Q., An, J., Xu, S., Xie, X., & Geng, J. (2021). Combining organic fertilizer with controlled-release urea to reduce nitrogen leaching and promote wheat yields. *Frontiers in Plant Science, 12*, 802137. https://doi.org/10.3389/fpls.2021.802137

Yang, Y. J., Lei, T., Du, W., Liang, C. L., Li, H. D., & Lv, J. L. (2020). Substituting chemical fertilizer nitrogen with organic manure and comparing their nitrogen use efficiency and winter wheat yield. *Journal of Agricultural Science, 158*, 262–268.

Yang, C., Huang, G., Chai, Q., & Luo, Z. (2011). Water use and yield of wheat/maize intercropping under alternate irrigation in the oasis field of northwest China. *Field Crops Research, 124*, 426–432.

Yu, H., Zhang, Q., Sun, P., & Song, C. (2018). Impact of droughts on winter wheat yield in different growth stages during 2001–2016 in Eastern China. *International Journal of Disaster Risk Science, 9*(3), 376–391. https://doi.org/10.1007/s13753-018-0187-4

Yu, Y., Stomph, T. J., Makowski, D., & Van der Werf, W. (2015). Temporal niche differentiation increases the land equivalent ratio of annual intercrops: A meta-analysis. *Field Crops Research, 184*, 133–144.

Zafar, J., & Muhammad, F. C. (2007). Effects of soil and foliar application of different concentrations of NPK and foliar application of $(NH_4)_2$ SO_4 on growth and yield attributes in wheat (Triticum aestivum L.). *Pakistan Journal of Plant Science, 13*(2), 119–128.

Zampieri, M., Ceglar, A., Dentener, F., & Toreti, A. (2017). Wheat yield loss attributable to heat waves, drought and water excess at the global, national and subnational scales. *Environmental Research Letters, 12*, 064008. https://doi.org/10.1088/1748-9326/aa723b

Zeng, L. (2005). Exploration of relationships between physiological parameters and growth performance of rice (Oryza sativa L.) seedlings under salinity stress using multivariate analysis. *Plant and Soil, 268*, 51–59.

Zhang, H., Yu, X., Jin, Z., Zheng, W., Zhai, B., & Li, Z. (2017a). Improving grain yield and water use efficiency of winter wheat through a combination of manure and chemical nitrogen fertilizer on the Loess plateau, China. *Journal of Soil Science and Plant Nutrition, 17*(2). https://doi.org/10.4067/S0718-95162017005000033

Zhang, P., Liu, Y., Liu, W., Cao, M., Massart, S., & Wang, X. (2017b). Identification, Characterization and full-Length sequence analysis of a novel polerovirus associated with wheat leaf yellowing disease. *Frontiers in Microbiology, 8*, 1689.

Zhang, F., Niu, J., Zhang, W., Chen, X., Li, C., Yuan, L., & Xie, J. (2010). Potassium nutrition of crops under varied regimes of nitrogen supply. *Plant and Soil, 335*, 21–34.

Zhao, X., Li, F., Ai, Z., Li, J., & Gu, C. (2018). Stable isotope evidences for identifying crop water uptake in a typical winter wheat-summer maize rotation field in the North China Plain. *Science of the Total Environment, 618*, 121–131. https://doi.org/10.1016/j.scitotenv.2017.10.315

Zhu, X. G., Long, S. P., & Ort, D. R. (2010). Improving photosynthetic efficiency for greater yield. *Annual Review of Plant Biology, 61*, 235–261.

Zia-ul-Hassan, Kubar, K. A., Tunio, I. R., Shah, A. N., Shah, J. A., & Maitlo, A. A. (2014). Evaluating potassium-use-efficiency of five cotton genotypes of Pakistan. *Pakistan Journal of Botany, 46*(4), 1237–1242.

Chapter 2
Increasing Land and Water Use Efficiencies of Wheat: Case Study of Egypt

Abstract Wheat is the most important food crop in Egypt as it is used in bread making. However, the amounts of wheat produced every year are not enough to meet the total local consumption of the Egyptian. In this chapter, we reviewed the Egyptian breeding efforts in producing resilient wheat cultivars. It has been reported that 50 wheat cultivars (bread and durum) were released between 1947 and 2012, where they are tolerant to different biotic and abiotic stresses. For example, Giza-168 cultivar is resistant to rusts, as well as tolerating water and heat stresses and Gemmeiza-12 cultivar is resistant to rusts and tolerate to heat stress. Several improved management practices to increase wheat production were reviewed, namely reducing the reliance on chemical fertilizer in wheat production by application of organic fertilizer, legume crops preceding wheat, and wheat intercropping systems with legume crops. Moreover, the advantage of intercropping fahl clover with wheat was presented in a field study, where fahl clover was intercropped with wheat, in three planting densities (PD) (15, 25, and 35%, PD) under three cultivation methods (basins, furrows and raised beds). The results indicated that the highest wheat yield was obtained with 15% fahl clover PD intercropped with wheat on raised beds. The highest values of water consumptive use, soil available nitrogen and percentage of organic matter were obtained with 35% fahl clover PD intercropped with wheat on raised beds. The highest values of land and water equivalent ratios were obtained with 25% fahl clover PD intercropped with wheat on raised beds. Thus, inclusion of legume crop in wheat-base cropping system by intercropping could attain the sustainable use of land by increasing soil available nitrogen and organic matter. It attained the sustainable use of water by using the applied water to wheat to produce both fahl clover and wheat. Furthermore, it increased land productivity through increasing wheat yield, compared to wheat sole cultivation and produce two crops on the same piece of land.

Keywords Resilient wheat cultivars · Improved management practices · Organic fertilizer · Legumes preceding wheat · Intercropping a legume crop with wheat · Fahl clover

S. Ouda and A. E. Zohry, *Integration of Legume Crops with Cereal Crops Under Changing Climate*, https://doi.org/10.1007/978-3-031-68102-8_2

Introduction

Wheat is the most important food crop in Egypt as it is used in bread making. The cultivated area of wheat is the highest among the winter crops, where wheat competes with Egyptian clover and most recently with sugar beet (Zohry & Ouda, 2019). The highest wheat cultivated area is located in the Nile Delta represents 57% of its total area, whereas smaller areas in Middle and Upper Egypt with about 18 and 17%, respectively. Wheat is also cultivated in the new reclaimed areas of Egypt, with an area of 8% of its total cultivated area (Zohry & Ouda, 2020). The total cultivated area of wheat was estimated by 1.42 million hectares in 2022, which produced 9.62 million tons of wheat (CAPMAS, 2022). These amounts of wheat are not enough to meet the total local consumption of the Egyptian population, where Egypt has the highest per capita consumption of wheat in the world. In 1960s, the annual per capita consumption rate in Egypt was less than 110 kg of wheat. The annual per capita consumption was increased in the 1980s to reach 175 kg. Whereas, in 2012, the annual per capita consumption was increase to reach 200 kg of wheat (Aegic, 2015).

As a result, the government of Egypt imports the rest of the required wheat amounts mainly from Russia, and Ukraine, (82% combined), in addition to other countries, such as USA and Australia, which was the situation before the Russian aggression against Ukraine in 2021. In an attempt to compensate for the shortage in wheat importation due to the war, Egypt bought wheat from India and France (USDA, 2022).

Figure 2.1 presents the progress in wheat production and imports amounts between 2010–2020

In an effort to overcome the increasing wheat production-consumption gap in Egypt, breeding programs have been implemented to produce new cultivars tolerant to biotic and abiotic stresses and can attain yield stability under these stressful conditions. The productivity of wheat was progressively increased in the last three decades

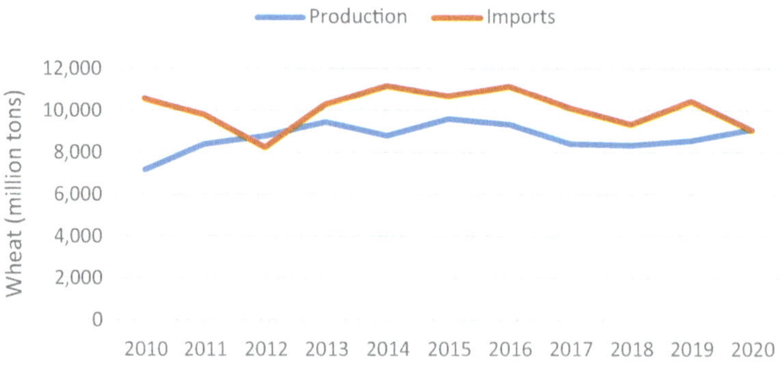

Fig. 2.1 Progress in wheat national production and imported amounts in the period of 2010–2020 in Egypt. *Data source* https://www.fao.org/faostat/en/#data

due to breeding efforts resulted in large varietal improvement (Abdelmageed et al., 2018).

Because agricultural practices influence the yield of wheat, the use of improved agricultural managements practices, or in other words, the use of climate-smart practices are the reason for the raise in wheat productivity. Examples of these practices is the precise land leveling, where it has been reported that it increases water use efficiency of wheat and save a reasonable amount of the applied irrigation water (Bahnas et al., 2010). Furthermore, cultivation of wheat on raised beds under surface irrigation system has many benefits. Swelam (2016) reported that the implementation of raised beds cultivation under surface irrigation results in 25% saving in the applied irrigation water to wheat, increased the yield by 30% and water use efficiency improved by 74%. Other researchers in Egypt explained the reason for yield increase under raised beds cultivation by the increase in the radiation use efficiency, which increase photosynthesis and consequently the final yield (Abouelenein et al., 2009; Zohry et al., 2020). Improvement in fertilizer use efficiency by wheat plants was also reported when raised beds cultivation of wheat was implemented (Khalil & Abouenein, 2012).

Matching local climatic conditions with plant growth and development process to define the suitable sowing date is also one of the improved management practices that was used to increase wheat productivity (Abdelmageed et al., 2018). Furthermore, defining the type and quantity of required amounts of fertilizers was a contributor in improving wheat productivity (Rekaby et al., 2016).

Figure 2.2 presents the progress in national wheat productivity in the period 1986–2019, which is due to breeding programs and using improved management practices.

One important approach to increase land and water use efficiency of wheat is to implement intercropping systems of wheat with other suitable crops on raised beds (Ouda & Zohry, 2017). Intercropping is an agricultural technology that involves planting two or more crop species together, where land productivity is increased (harvest two types of crops from the same area) (Porto et al., 2011), water productivity is increased (using the applied amount of irrigation water to one crop to irrigate two crops) and farmers income is increased (reduce risks from crop failure)

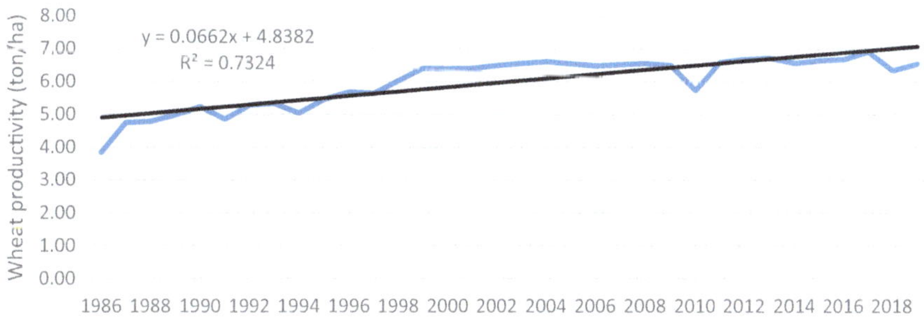

Fig. 2.2 Progress in wheat national productivity in the period of 1986–2019 in Egypt. *Data source* CAPMAS (1986–2018)

(Andersen, 2005). Intercropping legume crops with wheat not only increases its yield (Zohry et al., 2020), but also it increases nutrients availability (Li et al., 2014), and improves soil quality (Cong et al., 2015). Hamada and Hamd-Alla (2019) indicated that rotating wheat with legume crops and/or intercropping with legume crops can be advantageous in increasing wheat yield and improve the efficiency of water and fertilizer use.

In this chapter we reviewed the Egyptian breeding efforts in producing resilient wheat cultivars that can tolerate biotic and abiotic stresses. The improved management practices that can sustainably increase wheat productivity were also reviewed. Moreover, the advantage of intercropping fahl clover with wheat, as an example of intercropping a legume crop with wheat, was also presented.

Breeding Efforts to Produce Resilient Wheat Cultivars to Biotic and Abiotic Stresses

During wheat growing season, it exposed to external stress condition, namely biotic and abiotic stresses, which negatively impact its yield. Biotic stresses cause a great threat to wheat production globally. It is a result of several living organisms namely fungi (Tembo et al., 2020), virus (Zhang et al., 2017), insects (Katare et al., 2015), nematodes (Dababat et al., 2015), and weeds (Bertholdsson, 2005). Whereas, abiotic stress is resulted from drought, salinity and heat stress. These stresses reduce the potential yield performance of wheat cultivars to a massive extent.

Wheat Cultivars Resistant to Biotic Stress

Breeding for biotic stress resistance have released several wheat cultivars in Egypt. Aphid is one of the most severe pests attacking wheat causing damage to the plants either directly by sucking juice or indirectly as a vector of diseases (Mansour, 2012). El-Dessouki et al. (2022) revealed that Misr-1 cultivar was more resistant to the aphid species than Misr-3 cultivars. Furthermore, Megahed and Kamel (2020) revealed that Sids-12 cultivar is resistant to greenbug, while Giza-168 and Misr-1 cultivars were moderately resistant cultivars. On the contrary, BeniSuef-5 and Gimmiza-11 cultivars were susceptible to greenbug infestation. Whereas, Helmi and Rashwan (2013) indicated that Gemiza-9 appeared to be a resistant cultivar to aphid infestation.

Fungi infection is seed-borne diseases, which not only reduce wheat yield, but also cause deterioration in the marketable grain quality. Therefore, early detection of it is a crucial step in the diagnosis, as well as for the management programs in wheat (Majumder et al., 2013). Baka (2014) studied 14 Egyptian wheat cultivars with three seed-borne fungi were incident in them. He used the antifungal of water plant extract of plant Asclepias sinaica collected from the Sinai Peninsula, in which exhibited the most antifungal activity. Among the tested wheat cultivars, he found BaniSuef-5 cultivar has the minimum infested grain germination, in addition to attaining the maximum length of roots and shoots of the seedlings.

Furthermore, a fungal disease affected wheat has break lately in Egypt, known as Stemphylium leaf spot that caused by Stemphylium vesicarium Wallr (Simmons). Farag et al. (2022) reported that Sakha-95 cultivar showed low percentage of infection to it, which it can be rated as a resistant cultivar.

El-Shamy et al. (2012) indicated that wheat powdery mildew can cause severe damage to wheat, where losses in wheat yield due to powdery mildew may reach 10–18%. Abdelrhim et al. (2018) reported that two durum wheat cultivars, namely BeniSuef-5 and BeniSuef-6 showed partial resistance to powdery mildew.

Loose smut is among the most common and widespread diseases of wheat in Egypt (Esmail et al., 2021). Draz et al. (2021) assessed the incidence of loose smut disease in ten wheat cultivars and found that Misr-3 cultivar exhibits resistance, as a result of the presence of resistant gene.

Furthermore, yellow rust attacks most of Egypt's commercial wheat cultivars causing serious infections and high yield losses, especially in Northern governorates of Egypt (Omara et al., 2016). Mabrouk et al. (2022) reported that wheat cultivars Shandweel-1 and Sids-1 were moderately susceptible to yellow rust infection, whereas Sakha-94 cultivar recorded the lowest value of infection. Whereas, Hafez et al. (2022) reported that Sakha-62, Sakha-9, Sakha-95, Gemmiza-7, Sids-14 and Misr-3 cultivars showed high level of resistance to yellow rust. Gebrel et al. (2019) studied the relationship between nitrogen fertilizer and rust infection and severity in three wheat cultivars. They found that Giza-168 and Shandaweel-1 cultivars were more resistant to rust diseases followed by Gemmeiza-11 cultivar. Increased levels of nitrogen fertilizer than the recommended dose increased the severity of rust diseases during grain filling. Mabrouk et al. (2022) tested five wheat cultivars, namely Shandweel-1, Gemmeiza-11, Sids-1, Sids-12, Sakha-94 to yellow rust and found that the highly susceptible wheat cultivars were Sids-12 and Gemmeiza-11, the moderately susceptible cultivars were Shandweel-1 and Sids-1 and the resistant cultivar was Sakha-94.

Furthermore, leaf rust is a vital disease that cause significant yield losses in wheat fields. Thabet and Najeeb (2017) revealed that Sids-1 and Gimmeza-9 cultivars have high level of tolerance to leaf rust infection under the Egyptian field conditions. Atia et al. (2021) developed effective marker-assisted selection programs for leaf rust resistance in wheat mainly through the availability of deep mining of resistance genes within the germplasm collections. They used 50 Egyptian wheat cultivars at the adult plant stage for two successive seasons and identified the absence/presence of 28 leaf rust resistance (*Lr*) genes and were able to identify 21 *Lr* genes.

Wheat Cultivars Resistant to Abiotic Stress

Increasing the resilience of wheat cultivars towards various abiotic stresses is a promising method for maximizing crop production under adverse conditions of drought, salinity and heat stresses (Shen et al., 2015). Areas under drought are very prone to soil erosion, which in turns leads to reduced soil fertility, soil nutrients, and soil productivity, where all are important factors driving crop growth and productivity (Delgado et al., 2013). Abdelmageed et al. (2018) indicated that the breeding programs in Egypt were successful to release cultivars with shorter plant height to resist lodging, and cultivars with long spikes and large number of tillers to obtain high yield. The breeding was also done for improving nitrogen use efficiency, as well as increasing tolerance to water, heat and salinity stresses.

It has been reported that frequent droughts could increase the risk of wheat yield losses by almost 12% (Leng & Hall, 2019). Emam et al. (2022) conducted an investigation for genetic characterization and determination of drought tolerance of ten Egyptian cultivars of wheat and they found that three cultivars were drought-tolerant, namely Giza-171, Misr-2, and Giza-168. Whereas, Abdelghany et al. (2016) indicated that Sids-13 and Misr-1 cultivars produced the highest values of grain yield under water stress and can be identified as suitable genotypes for dry land areas, where drought stress more frequently occurs during post-anthesis stage. Heiba et al. (2021) stated that wheat cultivars Sids-12, Misr-2 and Gemmieza-9 are tolerant to water deficit. Moreover. Kandil et al. (2013) indicated that Sakha-93 and Gemmieza-7 cultivars surpassed Giza-168 cultivar under water stress in the most of the characteristics.

Wheat is known to be sensitive to salinity, which hinders the growth and development of the plants, leads to low or no productivity under extreme severity of salinity (Nassar et al., 2020). Ragab and Kheir (2019) tested four wheat cultivars, namely Shandweel-1, Gemmiza-10, Sakha-93 and Misr-2 for its tolerance to salinity stress and found that they all are salinity tolerance cultivars. They concluded that these four cultivars can be used as a source of improving salinity tolerance in the wheat breeding program. Gadallah et al. (2017) studied salinity stress tolerance of sixteen Egyptian local wheat cultivars, using two salinity levels, namely 4000 and 8000 ppm. They found that Sakha-93 and Sahel-1 cultivars were the most tolerant cultivars. Furthermore, the cultivars Sids-1, Shandweel-1, Misr-2 and Misr-1 were moderately tolerant to salinity stress. Whereas, Kandil et al. (2013) indicated that Sakha-93 and Gemmieza-7 cultivars surpassed Giza-168 cultivar under salinity in the most of the characteristics.

Heat stress negatively affects various plant development processes and eventually resulting in great yield losses (Grant et al., 2011). With respect to heat stress tolerance, Abou and Tabl (2014) identified two tolerant wheat cultivars to heat stress during germination stage, namely Sakha-93 and Gemmeiza-9. Whereas, Zakaria (2020) indicated that Misr-2, Sids-4, Sids-1, Shandaweel-1, and Sids-12 may perform as potential donor for heat tolerance traits in breeding programs. Shenoda et al. (2021) stated that the most tolerant wheat cultivars to heat stress were Masr-2, Sids-1, and

Giza-171. Whereas, Hamada et al. (2022) conducted a climate change simulation study to determine the suitable wheat cultivars that can tolerate future harsh conditions of climate change and they concluded that cultivars Giza-160 and Sakha-94 are suitable for growing in zones with harsh environments, such as the eastern desert and south Egypt.

Abdelmageed et al. (2018) listed 50 wheat cultivars (bread and durum) released between 1947 and 2012. These cultivars are tolerant to different biotic and abiotic stresses. For example, Giza-168 cultivar is resistant to rusts, as well as tolerating water and heat stresses. Gemmeiza-12 cultivar is resistant to rusts and tolerate to heat stress. Gemmeiza-10 and Gemmiza-11 cultivars are resistant to rusts only. Whereas, Sakha-93 and Sakha-94 cultivars are resistant to rusts, and tolerate to salinity and heat stresses. Both Misr-1 and Misr-2 cultivars are resistant to rusts, as well as water deficit, heat, and give high yield as a result of more tillers.

Improved Management Practices to Sustainably Increase Wheat Production

Reducing the Reliance on Chemical Fertilizer

Recently, there is uncareful use of chemical fertilizers causing hazardous effects on the environment. The use of chemical fertilizers should not threaten the environmental sustainability, while maintaining reasonable food production. It has been reported that the excessive application of chemical fertilizers threatened soil quality and caused soil nutrient imbalance, soil acidity, as well as soil pollution (Yang et al., 2018). Application of organic fertilizer sources could overcome the setback of the excessive use of chemical fertilizer on the environment. Organic fertilizer can provide the required plant nutrients, increase soil water holding capacity, as well as improving the fertility of the soil (Wick et al., 2017). Furthermore, inclusion of legume crops in the cropping system, either by rotating with wheat or intercropping with wheat can be another approach to reduce the reliance on chemical fertilizer.

Application of Organic Fertilizer

Application of organic fertilizer as an alternative to mineral nitrogen fertilizers can be considered as improved management practice to attain a more sustainable wheat production (Bezabeh et al., 2022). Application of farmyard manure to the soil improves its physical and chemical conditions year after year (Agamy et al., 2012). However, the use of organic fertilizer alone was not sufficient to meet the plant requirements of nutrients, thus it should be used as partial substitute to chemical fertilizer, where its use will reduce production costs and environmental pollution (Hossam El-Din, 2007). Application of farmyard manure and bio-fertilizer to replace

50% of mineral fertilizer of wheat increase wheat yield by an average of 44% over two growing seasons (Attia & Abd El Salam, 2016).

Whereas, poultry manure has long been recognized as the most desirable organic fertilizer because it improves soil fertility by adding both major and essential nutrients, as well as soil organic matter, which improve moisture and nutrient retention (Farhad et al., 2009). Abd El-Samie et al. (2018) indicated that application of 24 m^3/ha of poultry manure to wheat plants and foliar spraying with mixture micronutrients, as well as addition of 192 kg N/ha improved the productivity of wheat yield under sandy soil condition in Egypt.

Moreover, composting is an efficient, cost-effective, and environmentally safe biological process for the recycling of residual agricultural biomasses (Pane et al., 2015). Using compost as fertilizer increases soil organic content, microbial activity (Scotti et al., 2015) and concentration of plant nutrients (Donn et al., 2014). Application of compost in combination with mineral fertilizer had a positive effect on wheat yield, compared to application of mineral fertilizer only (Ibrahim et al., 2012). Furthermore, Sorour et al. (2019) reported that application of 50% of chemical fertilizer and 50% compost to wheat increased its yield by 9% under the condition of salt-affected soil. Whereas, in sandy soil, application of compost increased wheat yield by 9%, whereas application of compost and biofertilizer increased yield by 33%, compared to application of chemical fertilizer only (Mohamed et al., 2019).

Legume Crops Preceding Wheat

Cereal crops generally have higher yields when cultivated after legume crops (Preissel et al., 2015), which is explained by nitrogen benefits from the legume crop residue. Guinet et al. (2020) indicated that wheat yield of unfertilized plants was higher by 27 and 25% after faba bean and lentil, respectively compared to the wheat yield obtained after barley. Furthermore, wheat yield was higher by 66 and 51% after summer legumes, compared to the wheat yield obtained after sorghum.

Fahl clover is a mono-cut variety of Egyptian clover, which has stem branching ability and is cut only once. Fahl clover is characterized by rapid growth and large forage yield, and it can be cultivated as early winter crop in September and grow until the beginning of November, where its growing season between 60 and 70 days (Bakheit et al., 2016). Therefore, it can precede winter crops to improve soil nitrogen content and improve the yield of the following winter crops.

Awad et al. (2023) indicated that soil nitrogen, phosphorous and potassium were increased, when fahl clover or intercropping system with a legume crop preceded wheat, compared to maize preceded wheat. They also stated that cultivating fahl clover in September and harvest it two weeks before wheat cultivation, increased wheat yield by 8%. Whereas, cultivating soybean intercropping system with maize in the summer and harvest it before wheat cultivation resulted in an increase in wheat yield by 26%. Additionally, cultivating cowpea intercropping system with maize, followed by fahl clover before wheat cultivation, resulted in the highest wheat yield, namely 20%.

Furthermore, Zohry et al. (2017a) indicated that inclusion of fahl clover between maize and wheat resulted in an increase in wheat yield by 32% and increase in water productivity of the system by 24%. Furthermore, cultivation of cowpea in the summer, fahl clover as an early winter crop and then wheat, highly increased wheat yield by 55% and increase water productivity of the system by 66%. Whereas, cultivation of cowpea intercropping system with maize, fahl clover and then wheat also increased wheat yield with lower percentage than the previous sequence, namely 42% and increase water productivity of the system by 17%. In another investigation by Ouda et al. (2020), the effect of the different crop sequences was studied, with respect to the applied water to each crop sequence, wheat yield and soil nitrogen percentage. The results indicated that the applied water to wheat was reduced by 16, 18 and 19% in the crop sequences: maize, fahl clover and then wheat; cowpea, fahl clover and then wheat; and cowpea intercropping system with maize, fahl clover and then wheat, respectively (scheduling irrigation and fertigation were implemented), compared to maize cultivation followed by wheat (fertilizer was broadcast on the soil before irrigation). Whereas, wheat yield was increased by 30, 55, and 43%, for the above crop sequences, respectively compared to wheat yield obtained from maize followed by wheat. Furthermore, soil nitrogen content was respectively increased by 50, 101, and 69%, for the above crop sequences, respectively compared to its value after maize followed by wheat.

Changing crop sequence from rice then wheat to rice, fahl clover and then wheat resulted in an increase in wheat yield by 21% (Hamdany et al., 2016). Another research reported an increase in wheat yield by 12% when crop sequence was changed from rice then wheat to rice, fahl clover and then wheat (Melad, 2018). Hamada and Hamd-Alla (2019) indicated that fahl clover cultivation before wheat increase wheat yield by 6%, compared to fallow preceding wheat.

Wheat Intercropping Systems with Legume Crops

Wheat intercropping systems with legume crops could play an important role in the sustainable use of soil and water resources. Zohry et al. (2017b) intercropped pea with wheat (100% pea + 80% wheat of its recommended planting density) and they found that wheat yield was increased by 4%, compared to the yield obtained from sole planting of 80% wheat. Whereas, Abbady et al. (2016) obtained similar percentage of increase in wheat yield and an increased in soil organic matter by 17%, compared to the value obtained under sole wheat cultivation. Furthermore, Zohry et al. (2020) reported that intercropping pea with wheat (50% pea + 100% wheat of its recommended planting density) resulted in an increase in wheat yield by 5% in the second growing season, compared to the yield of sole wheat. Furthermore, soil organic matter was increased after the harvest in the second season by 11%, compared to its value before cultivation. The results also showed saving in the applied irrigation water to this intercropping system by 23%, compared to the value of applied water to the sole wheat planted by the farmer. A successful wheat intercropping system with fahl clover was implemented by Ali (2018) where wheat yield was increased by 7% when fahl clover was intercropped with wheat on raised beds compared to planting wheat drill in basins cultivation and by 17%, compared to broadcast wheat seeds in basins.

Furthermore, a reduction in wheat yield under intercropping system with a legume crop was reported by other researchers. Abd-Rabboh and Koriem (2021) implemented an intercropping system of pea with wheat, where wheat planting density was 100% and pea planting density was 50%. They concluded that wheat yield was reduced by less than 1%, compared to yield value obtained under sole wheat planting. Similar results were obtained by Sheha et al. (2015), where they intercropped pea with wheat and wheat yield was reduced by less than 1%, compared to yield value obtained under sole cropping of wheat.

Furthermore, Abou-Keriasha et al. (2013) reported that faba bean intercropping system with wheat (50% faba bean + 100% wheat of its recommended planting density) reduced wheat plant height, and number of spikes per m^2. Consequently, grain yield was reduced by 8%, compared to the yield obtained under the sole planting of wheat. Similar trend was obtained for faba bean under its intercropping system with wheat, where plant height, number of branches per plant and seed yield were also reduced, compared to the yield obtained under faba bean sole planting. These results implied the high competition between the two crops, which negatively affected the yield of both crops. Similarly, Hamada and Hamd-Alla (2019) reported a reduction in wheat yield under its intercropping system with faba bean (50% faba bean + 100% wheat of its recommended planting density) by 3%, compared to the yield obtained under sole planting. However, when fahl clover preceded faba bean intercropping system with wheat, wheat yield was increased by 2%, compared to the yield obtained under its sole planting.

Fehl Clover Intercropped with Wheat: A Field Study

Fahl clover is a short-season (60–70 days) single-cut clover (a legume crop) providing high dry matter green forage suitable for hay production. To produce its seeds, a part of its cultivating area is devoted for that. Another method to produce fahl clover seeds is to intercrop it with wheat, where a low percentage of its seeds can be mixed with wheat grains and cultivated together in the same sowing day of wheat and harvested in the same day. Using a sieve, the seeds of both crops are easily separated. Intercropping fahl clover with wheat has many benefits to the soil and both crops use the applied irrigation water to wheat.

To determine the effect of intercropping fahl clover with wheat on land and water use efficiency, two field experiments were conducted during the two successive growing seasons of 2019/20 and 2020/21 at Assuit Governorate (27.1783° N, 31.1859° E), Upper Egypt. The experimental area has an arid climate with cool winters and hot, dry summers. The data of average monthly solar radiation, and air temperature (maximum, minimum) and dew point during the studied growing seasons were obtained from the following website: https://power.larc.nasa.gov and used to calculate reference evapotranspiration (ETo) using the Penman–Monteith equation (Allen et al., 1998) (Fig. 2.3a, b).

The soil of the experimental site is classified as clayey. The physical and chemical properties of the soil were done according to Klute (1986) and Tan (1996) (Tables 2.1 and 2.2).

The tested treatments included the interaction between intercropping of three planting densities of fahl clover (15, 25, and 35% of its recommended planting density), with wheat (100% of its recommended planting density) and three cultivation methods (broadcast in basins (control), furrows, and raised beds). The aim of the experiment was to test which wheat intercropping system with fahl clover under the three cultivation methods attains the highest land and water use efficiencies, as well as the highest soil nitrogen content and organic matter. The experiment was implemented in a split-plot design with three replications. The main plots were assigned to cultivation methods and the subplots were assigned to wheat intercropping systems with fahl clover. Fahl clover seeding rate was 15, 25 and 35% of its recommended rate of 48 kg/ha. Each rate was mixed with the recommended rate of wheat seeding rate of 120 kg/ha. In addition, cultivation of sole wheat and sole fahl clover were also implemented.

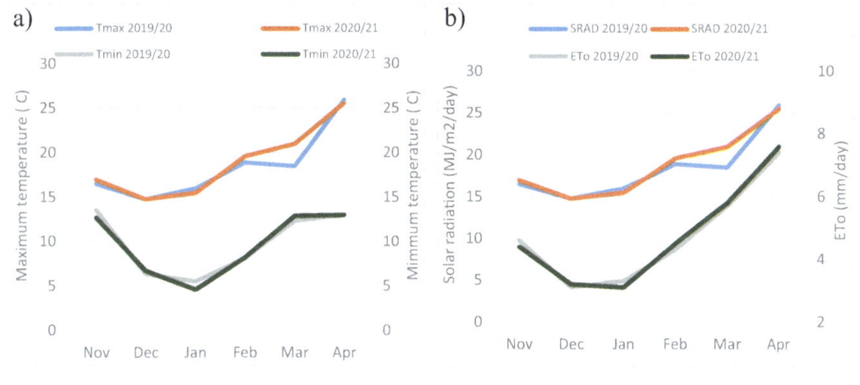

Fig. 2.3 Averaged monthly maximum temperature (T_{max}), and minimum temperature (T_{min}) (**a**), solar radiation (SRAD) and ETo values (**b**) for the experimental site in both growing seasons

Table 2.1 Main physical properties of soil at the experimental site before cultivation

Soil depth (cm)	Particle size distribution			Texture class	Bulk density (g m^{-3})	Field capacity (%)	Available water (%)	Permanent wilting point (%)
	Sand (%)	Silt (%)	Clay (%)					
0–20	12.32	23.15	65.24	Clayey	1.07	45.16	24.25	21.13
20–40	16.14	24.16	60.21		1.26	40.15	22.14	18.32
40–60	18.99	26.18	56.23		1.45	35.25	18.56	16.23

Table 2.2 Chemical properties and available macronutrients of the soil at the experimental site before cultivation

Properties	Chemical properties
pH (1:2.5)	7.66
EC (dS/m)	1.60
$CaCO_3$ (%)	3.88
Available N (ppm)	22.9
Available P_2O_5 (ppm)	9.0
Available K_2O (ppm)	350.0
Organic matter (%)	1.68

The land was prepared by ploughing twice and then was leveled. Wheat cultivar Sids-1 and fahl clover cultivar, either sole or intercropped were planted on the 20th of November and the 18th of November in the first and second seasons, respectively. Superphosphate (36 kg P_2O_5/ha) and potassium sulfate (57.6 kg K_2O/ha) were added during land preparation, as recommended by the Egyptian Ministry of Agriculture and Land Reclamations. Nitrogen fertilizer rate was applied in three equal doses. The first dose was added at planting date, the second dose was added at the first irrigation and the third dose added at the third irrigation. The plot area was 12 m^2 (3×4 m). The preceding crop was maize in both growing seasons, planted in May and harvest in the beginning of September.

Surface irrigation was used to deliver water to the studied cropping systems, where 5 irrigations were applied to either sole wheat, sole fahl clover, or fahl clover intercropping systems with wheat. Irrigation was applied as 120% ETo every 21 days. The values of water consumptive use (WCU, mm) were calculated using soil sampling methods according to the equation of Israelsen and Hansen (1962) as follows:

$$\mathrm{WCU} = \sum_{i=1}^{i=3} \frac{\theta_2 - \theta_1}{100} \times d \tag{2.1}$$

where: i = number of soil layer, θ_2 = soil moisture content after irrigation, (%, by volume), θ_1 = soil moisture content just before irrigation, (%, by volume), d = depth of soil layer (mm).

The amount of required irrigation water was calculated according to the following equation (Vermeiren & Jopling, 1984):

$$\mathrm{AIW} = \frac{\mathrm{ETo} \times I}{E_{\mathrm{a}}(1 - \mathrm{LR})} \tag{2.2}$$

where: AIW = applied irrigation water depth (mm/day), E_a = application efficiency (60% surface system). LR = leaching requirements (was not considered). We assumed that the applied irrigation water to sole wheat is enough to fulfill the required water for its intercropping systems with fahl clover.

Irrigation application efficiency was calculated under each cultivation method and intercropping system using the following equation (Allen et al., 1998).

$$AE(\%) = \frac{WCU}{AIW} \tag{2.3}$$

Other management practices, namely disease and pest control programs for wheat and fahl clover were followed as recommended by the Egyptian Ministry of Agriculture. Harvest took place on the 15th of May and the 11th of May in the first and second season, respectively, where the grain yield of wheat and the seeds of fahl clover were recorded based on the experimental plot area through harvesting all plants, separated the seeds and weighting them. Then, the weight of each crop in all the plots were combined to calculate yield per hectare.

Measurement of Soil Nitrogen and Organic Matter

A representative soil samples were collected before planting and after harvest of the second growing season to determine total soil nitrogen and organic matter. Using a hand-pushed auger, 12 cores or small samples were taken randomly within each treatment and replicated. Soil samples were air-dried, crushed gently and then passed through a 2 mm sieve to obtain uniform representative samples. Nitrogen was determined by the micro-Kjeldahl method (Hesse, 1971).

The carbon content was measured using wet oxidation (Walkley & Black, 1934). The organic matter content was calculated by multiplying organic carbon with 1.724.

Evaluation of the Studied Intercropping Systems

1. **Land equivalent ratio (LER)**: It is an evaluation of the land utilization efficiency of the intercropping system (Rao & Willey, 1980). It is calculated as follows:

$$LER = LER_A + LER_B = \left(Y_{int,A}/Y_{mono,A}\right) + \left(Y_{int,B}/Y_{mono,B}\right) \tag{2.4}$$

where: LER_A and LER_B are the partial land equivalent ratio of crop A and crop B, respectively. $Y_{int,A}$ and $Y_{int,B}$ are the intercropped yield of crop A and crop B, respectively. $Y_{mono,A}$ and $Y_{mono,B}$ are the monoculture yield of crop A and crop B, respectively. If LER > 1, it indicates high land utilization efficiency of the intercropping system than that of monoculture.

2. **Percentage of land saved (LS, %)**: It indicates the lands saved from intercropping that could be used for other agricultural purposes. It was calculated according to Rao and Willey (1980) as followed:

$$\mathrm{LS}(\%) = 100 - (1/\mathrm{LER}) \times 100 \tag{2.5}$$

3. **Water equivalent ratio (WER)**: It quantifies the amount of water that would be needed in single crops to achieve the same yield as produced with one unit of water in intercrops and it is calculated according to the formula of Mao et al. (2012) as follows:

$$\begin{aligned}\mathrm{WER} = &\left[\left(Y_{\mathrm{int,A}}/\mathrm{WU}_{\mathrm{int}}\right)/\left(Y_{\mathrm{mono,A}}/\mathrm{WU}_{\mathrm{mono,A}}\right)\right] \\ &+ \left[\left(Y_{\mathrm{int,B}}/\mathrm{WU}_{\mathrm{int}}\right)/\left(Y_{\mathrm{mono,B}}/\mathrm{WU}_{\mathrm{mono,B}}\right)\right]\end{aligned} \tag{2.6}$$

where: $Y_{\mathrm{int,A}}$, $Y_{\mathrm{mono,A}}$, $Y_{\mathrm{int,B}}$, and $Y_{\mathrm{mono,B}}$ are yield of intercropped crop A, yield of mono crop A, yield of intercropped Crop B and yield of mono crop B, respectively. $\mathrm{WU}_{\mathrm{int}}$, $\mathrm{WU}_{\mathrm{mono,A}}$, and $\mathrm{WU}_{\mathrm{mono,B}}$ are water use by the intercropping system, water use by mono Crop A, water use by intercropped crop B, respectively. If the WER > 1, it suggests that the production of unit of water is higher than that of monoculture.

Wheat Yield, Fahl Clover Yield and the Applied Irrigation Water Under Its Cropping Systems

The results in Table 2.3 indicated that, in the first growing season, sole wheat yield was higher than its value under intercropping systems with fahl clover. On the contrary, in the second growing season the sole wheat yield was lower than its value under its intercropping systems with fahl clover. This result was true under the three cultivation methods. Furthermore, both the yield of sole or intercropped wheat with fahl clover in the first season was lower than its values in the second season under the three cultivation methods. Similar results were obtained in Egypt by Ali et al. (2017), where they reported a reduction in the sole wheat yield in the second growing season than the first and increase in wheat yield under its intercropping systems with fahl clover, compared to the sole yield. Table 2.3 also showed that the yield of fahl clover increased as its planting density increased in both growing seasons and it was higher in the second growing season, compared to the first growing season and it was the highest under raised beds cultivation.

These results could be attributed to the residual effect of fahl clover as a legume crop in the second season. Kirihetti (2018) indicated that an amount of fixed nitrogen returned to the soil as crop residue was observed after legumes cultivation. Fahl clover has the ability to perform symbiotic nitrogen fixation in the nodules of its roots, which increases soil nitrogen availability for the intercropped crop (Ivanov et al., 2012). Furthermore, Mat Hassan et al. (2012) indicated that legumes mobilize phosphorus in the soil during its growth, which increases P uptake by the following cereal crop.

Furthermore, the highest yield of wheat either sole or intercropped was obtained under cultivation on raised beds (Table 2.3). It has been reported that cultivation

Table 2.3 Wheat yield (WY), fahl clover yield (FY) and the applied irrigation water (AIW) in both growing seasons

	Broadcast in basins			Furrows			Raised beds		
	WY (ton/ha)	FY (ton/ha)	AIW (mm)	WY (ton/ha)	FY (ton/ha)	AIW (mm)	WY (ton/ha)	FY (ton/ha)	AIW (mm)
First growing season									
SF	–	0.73	152	–	0.73	135	–	0.73	119
SW	6.96	–	652	7.47	–	588	7.88	–	460
FIW1	5.76	0.17	654	6.36	0.18	590	6.96	0.19	463
FIW2	5.70	0.23	659	6.27	0.24	593	6.82	0.27	464
FIW3	5.05	0.29	662	5.39	0.30	595	5.67	0.33	468
Second growing season									
SF	–	0.81	149	–	0.81	129	–	0.81	110
SW	6.52	–	645	7.25	–	583	7.28	–	458
FIW1	6.37	0.18	655	7.27	0.19	591	7.76	0.21	462
FIW2	6.30	0.24	658	7.23	0.25	595	7.26	0.27	463
FIW3	5.61	0.30	660	6.27	0.32	597	6.57	0.34	465

SF = sole fahl clover, SW = sole wheat, FIW1 = 15% fahl clover intercropped with 100% wheat, FIW2 = 25% fahl clover intercropped with 100% wheat, FIW3 = 35% fahl clover intercropped with 100% wheat

on raised beds can increase productivity, as a result of an increase in radiation used efficiency because crops are more exposed to solar radiation (Zohry & Ouda, 2019). Furthermore, Majeed et al. (2015) indicated that raised beds planting of wheat improved fertilizer use efficiency. It creates a better soil physical environment throughout the crop growth period (Aggarwal & Goswami, 2003). Cultivation on raised beds enhances root growth and increases root length density in the upper 45 cm in the raised beds due to porous soil environment (Dey et al., 2015).

Under the three cultivation methods, wheat yield was the highest under its intercropping system with the lowest fahl clover planting density (15% of its recommended density) in both growing seasons (Table 2.3), which could be explained by mutual shading between both crops the is the lowest. Nevertheless, the highest values of wheat yield under this system were obtained under raised beds cultivation, namely 6.96 and 7.76 ton/ha in the first and second growing season, respectively (Table 2.3). Similar results were reported by Abou-Kerisha et al. (2008, 2013), where they obtained an increase in wheat yield when intercropped with fahl clover under low fahl clover seeding rate.

Table 2.3 also showed that the applied irrigation water was gradually reduced under raised beds cultivation, compared to broadcast in basins and furrows. This result hold true in both growing seasons. It has been reported that cultivation on raised bed save on the applied irrigation water under surface irrigation system (Majeed et al., 2015). Furthermore, increasing the percentage of fahl clover from 15 to 25%, slightly increased the applied irrigation water, which is an indicator of efficient water use by the intercropping system.

Yield Increase and Water Saving Under Wheat Cropping Systems

Table 2.4 presents a comparison between the wheat yield resulted from cultivation on furrows or on raised beds with broadcasting in basins, as well as the saving in the applied irrigation water. The results in Table 2.4 indicated that the highest percentage of increased in wheat yield was attained under 15% fahl clover intercropped with 100% wheat on raised beds, namely 21 and 23% in the first and second growing season, respectively, compared to cultivating wheat in basins. The lowest value of percentage of increase in wheat yield was attained for sole wheat cultivated on furrows, namely 7 and 12%, in the first and second growing season, respectively, compared to cultivating wheat in basins. Sing et al. (2010) found that cultivation on raised beds gave higher wheat yield than its value obtained under conventional flat beds planting. It has been also reported that cultivation on raised beds can increase productivity by 15–22%, as a result of increase in fertilizer use efficiency (Abouelenein et al., 2010; Khalil and Abouenein, 2012; Zohry et al., 2020).

Table 2.4 also shows that a saving in the applied irrigation water occurred when cultivation method was change from broadcasting in basins to raised beds cultivation. The percentage of saving in applied irrigation water was 10% under furrows cultivation and increased to 29–30% under raised beds cultivation in both seasons. Sing et al. (2010) found lower water consumption and higher wheat yield under raised beds planting than that under conventional flat beds planting due to decrease in the applied irrigation amount. Karrou et al. (2012) stated that cultivation of wheat on raised beds

Table 2.4 The percentage of increase in wheat grain yield and the percentage of saving in the applied irrigation water as a result of cultivation method under wheat cropping systems

	Percentage of increase in wheat yield (%)			Percentage of saving in the applied irrigation water (%)		
	Broadcast in basins	Furrows	Raised beds	Broadcast in basins	Furrows	Raised beds
First growing season						
Sole	–	7	13	–	10	29
FIW1	–	11	21	–	10	29
FIW2	–	10	18	–	10	30
FIW3	–	7	12	–	10	29
Second growing season						
Sole	–	11	12	–	10	29
FIW1	–	14	23	–	10	30
FIW2	–	14	14	–	10	30
FIW3	–	12	17	–	10	30

FIW1 = 15% fahl clover intercropped with 100% wheat, FIW2 = 25% fahl clover intercropped with 100% wheat, FIW3 = 35% fahl clover intercropped with 100% wheat

resulted in a saving by 28% in the applied irrigation water, compared to farmer irrigation and resulted in a saving by 26% of the applied irrigation water, compared to furrow cultivation. Whereas, El-Hadidi et al. (2014) reported 17% saving in the applied irrigation under wheat cultivation on raised beds, compared to flat cultivation. Furthermore, Ahmad et al. (2009) reported that raised beds cultivation increased water use efficiency, where 20–25% of irrigation water was saved.

Thus, the results in Table 2.4 implied that the saving in the applied irrigation water was mainly due to changing cultivation method from broadcast in basins to raised beds cultivation, regardless of the intercropping system.

Water Consumptive Use of Wheat Cropping Systems

Figure 2.4 shows the water consumptive use of the sole wheat and its intercropping systems with fahl clover under the three cultivation methods in both growing seasons. The sole cultivation of wheat broadcasted in basins attained the lowest value of water consumptive use, whereas the highest value was obtained for 35% fahl clover intercropped with wheat under raised beds cultivation. Increasing the planting density of fahl clover in the intercropping system with wheat increased water consumptive use. This result was true in both growing seasons. Similar results were obtained by El-Shamy et al. (2023) and Ali et al. (2017), where they stated that the lowest planting density of fahl clover in its intercropping system with wheat consumed the lowest value of the applied irrigation water.

Our results showed that water consumption by wheat intercropping systems with fahl clover was higher than that of sole wheat (Fig. 2.4a, b). This result is supported by the findings of Morris and Garrity (1993) and Mao et al. (2012). Furthermore, Yin

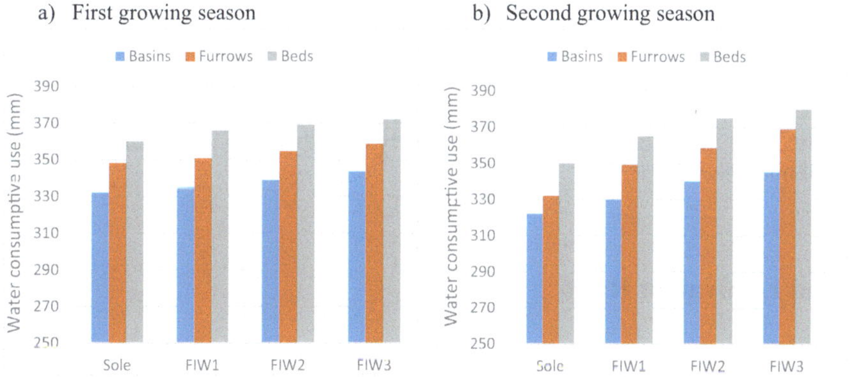

Fig. 2.4 Water consumptive use of the sole wheat and its intercropping systems with fahl clover under the three cultivation methods in both seasons. FIW1 = 15% fahl clover intercropped with 100% wheat, FIW2 = 25% fahl clover intercropped with 100% wheat, FIW3 = 35% fahl clover intercropped with 100% wheat

Table 2.5 Irrigation application efficiency for different cultivation method under wheat cropping systems

	First growing season			Second growing season		
	Broadcast in basins	Furrows	Raised beds	Broadcast in basins	Furrows	Raised beds
Sole	51	59	78	48	57	76
FIW1	51	59	79	50	59	79
FIW2	51	60	80	52	60	81
FIW3	52	60	79	52	62	82

FIW1 = 15% fahl clover intercropped with 100% wheat, FIW2 = 25% fahl clover intercropped with 100% wheat, FIW3 = 35% fahl clover intercropped with 100% wheat

et al. (2019) reported that daily soil evaporation of intercropping was less than that of sole cropping, indicating a significant advantage over sole cropping in improving crop water availability.

Irrigation Application Efficiency Under Wheat Cropping Systems

The irrigation application efficiency under each cultivation method and each fahl clover intercropping system with wheat was calculated and presented in Table 2.5. The results revealed that the lowest irrigation application efficiency was found under sole wheat broadcasted in basins in both growing seasons. Increasing fahl clover planting density under its intercropping system with wheat increased irrigation application efficiency, with the highest value was obtained under raised beds cultivation and 35% fahl clover intercropped with 100% wheat, namely 82% in both growing seasons. These results prove that cultivation on raised beds improved irrigation water use by the crops. Dogan and Kirnak (2010) stated that cultivation on raised beds exerts control on the applied water resulting in less water losses through deep percolation, thus reduced total applied water for the crops. Furthermore, Ahmad et al. (2009) reported that raised beds planting significantly contributed in improving soil water distribution and efficiency.

Land Equivalent Ratio

Land equivalent ratio (LER) is used as an indicator of the efficiency of the intercropping system in utilizing the cultivated area. LER describes the ratio of the area under sole cropping to the area under intercropping needed to the fractions of the intercropped give an equal amount of yields at the same management level. LER is calculated as the sum of yields divided by the sole crop yields (Mead & Willey, 1980).

The results in Table 2.6 showed that intercropping fahl clover with wheat had a positive effect on the dry matter production over the season reflected by high value of LER, which demonstrated an increase in land use efficiency. The value of land equivalent value of fahl clover (LER_F) increased as its planting density increases from 15 to 35%. All the values of LER_F were higher than its relative planting density. The highest values of LER_F were found under raised beds cultivation of 35% fahl clover intercropped with wheat, namely 0.44 and 0.42 in the first and second growing seasons, respectively. Whereas, the highest values of land equivalent ratio of wheat (LER_W) were found under raised beds cultivation 15% fahl clover intercropped with wheat in both growing season under raised beds, where it reached the value of 0.88 and 1.07 in the first and second growing seasons, respectively. This high value of LER_W attained as a result of the residual effect of fahl clover on the soil in second growing season, which positively reflected on the sole and intercropped wheat yield. The highest values of total land equivalent ratio (LER_{Total}) were obtained under raised beds cultivation of 25% fahl clover intercropped with wheat, namely 1.23 and 1.33 in the first and second growing season, respectively. This result can be explained by the reduction in wheat yield under its intercropping system with 35% fahl clover, which reflected on lower LER_{Total}, compared to its value under 25% fahl clover, which implied a significant yield advantage of 25% fahl clover intercropping system. Under this intercropping system, 23 to 33% more land would have been required if fahl clover was planted as sole crop to obtain the same yield as obtained in that intercropping system. Similar results were obtained by Ali (2018), who reported LER values of 1.25 when fahl clover was intercropped with wheat on raised beds.

Table 2.6 Land equivalent ratio (LER) for wheat intercropping systems with fahl clover under the studied cultivation methods in both growing seasons

	Broadcast in basins			Furrows			Raised beds		
	LER_F	LER_W	LER_{Total}	LER_F	LER_W	LER_{Total}	LER_F	LER_W	LER_{Total}
First growing season									
FIW1	0.23	0.83	1.06	0.25	0.85	1.10	0.26	0.88	1.15
FIW2	0.32	0.82	1.13	0.33	0.84	1.17	0.36	0.87	1.23
FIW3	0.40	0.73	1.12	0.41	0.72	1.13	0.44	0.72	1.16
Second growing season									
FIW1	0.22	0.98	1.20	0.24	1.00	1.24	0.25	1.07	1.32
FIW2	0.30	0.97	1.27	0.31	1.00	1.31	0.33	1.00	1.33
FIW3	0.37	0.86	1.23	0.40	0.86	1.26	0.42	0.90	1.32

LER_F = partial LER for fahl clover, LER_W = partial LER for wheat, LER_{Total} = LER of the total system, FIW1 = 15% fahl clover intercropped with 100% wheat, FIW2 = 25% fahl clover intercropped with 100% wheat, FIW3 = 35% fahl clover intercropped with 100% wheat

Table 2.7 Percentage of saved land under wheat intercropping systems with fahl clover under the studied cultivation methods in both growing seasons

	First growing season			Second growing season		
	Broadcast in basins	Furrows	Raised beds	Broadcast in basins	Furrows	Raised beds
FIW1	5.64	9.07	12.88	16.57	19.28	24.26
FIW2	11.85	14.55	18.53	21.23	23.77	24.81
FIW3	10.93	11.62	13.96	18.74	20.80	24.16

FIW1 = 15% fahl clover intercropped with 100% wheat, FIW2 = 25% fahl clover intercropped with 100% wheat, FIW3 = 35% fahl clover intercropped with 100% wheat

Percentage of the Saved Land

Implementing intercropping systems resulted in saving in the cultivated area due to the sharing of two crops the same piece of land. The results in Table 2.7 indicated that the percentage of saved land was higher in the second growing season than the first growing season. The highest values of the percentage of saved land were found under raised beds cultivation of 25% fahl clover intercropped with wheat, namely 18.53 and 24.81% in the first and second growing season, respectively. This result implied that it could cost about 0.185 and 0.248 hectare in the first and second growing season, respectively if a sole fahl clover was planted to get the same yield that harvested from 1.0 hectare through this intercropping system.

Water Equivalent Ratio

Water equivalent ratio (WER) is used to determine whether irrigation water was used more efficiently in the intercropping system than in the sole cultivation. It has been reported that intercropping considerably enhances water use efficiency, which is affected by the sowing proportion of the intercropping system, and that reflecting high water use advantage (Mao et al., 2012). Additionally, intercropping with legumes play a role in improving soil characteristics, such as increasing soil aggregate stability (Rücknagel et al., 2016), as well as water holding and infiltration (Wick et al., 2017).

The results in Table 2.8 revealed that the water equivalent ratio values of fahl clover (WER_{F}) were lower in the second season, compared to the first season, whereas the values of water equivalent ratio value for wheat (WER_{W}) were higher in the second season, compared to the first season. WER_{F} increased as its planting density increases from 15 to 35%. The highest values of WER_{F} were found under raised beds cultivation of 35% fahl clover intercropped with wheat, namely 0.45 and 0.42 in the first and second growing seasons, respectively. Whereas, the highest value of WER_{W} were found under raised beds cultivation in the second growing season under 15% fahl clover intercropped with wheat, where it was 0.87 and 1.00 in the first and second

Table 2.8 Water equivalent ratio (WER) for wheat intercropping systems with fahl clover under the studied cultivation methods in both growing seasons

	Broadcast in basins			Furrows			Raised beds		
	WER_F	WER_W	WER_{Total}	WER_F	WER_W	WER_{Total}	WER_F	WER_W	WER_{Total}
First growing season									
FIW1	0.23	0.82	1.05	0.25	0.85	1.09	0.27	0.87	1.14
FIW2	0.32	0.80	1.12	0.33	0.82	1.16	0.36	0.84	1.21
FIW3	0.40	0.70	1.10	0.42	0.70	1.11	0.45	0.74	1.19
Second growing season									
FIW1	0.22	0.89	1.11	0.24	0.96	1.19	0.25	1.00	1.25
FIW2	0.30	0.86	1.16	0.31	0.93	1.24	0.33	0.93	1.26
FIW3	0.25	0.83	1.09	0.33	0.78	1.11	0.42	0.82	1.23

FIW1 = 15% fahl clover intercropped with 100% wheat, FIW2 = 25% fahl clover intercropped with 100% wheat, FIW3 = 35% fahl clover intercropped with 100% wheat

growing season, respectively. This high value of WER_W attained as a result of the residual effect of fahl clover on the soil in second growing season, which increase soil water holding capacity and that positively reflected on the intercropped wheat yield.

The highest values of total water equivalent ratio (WER_{Total}) were obtained under raised beds cultivation of 25% fahl clover intercropped with wheat, namely 1.21 and 1.26 in the first and second season, respectively. This result implied an advantage of this intercropping system over the other studied wheat intercropping systems. Thus, enhancement in water use efficiency by 21 to 26% can obtained if this intercropping system was implemented (Table 2.8).

Wang et al. (2018) indicated that the appropriate planting density improving photosynthesis, microclimate environment between the two intercrops, increasing root length density and root absorption area.

Soil Available Nitrogen and Percentage of Organic Matter

Intercropping with legumes allows the production of high yields with lower inputs of nitrogen fertilizer (Li et al., 2020). Symbiotic nitrogen fixation performed in the nodules of legume roots is a key agro-ecosystem service provided by legumes, which increases soil nitrogen availability for intercropped crops (Ivanov et al., 2012). Dwivedi et al. (2015) reported that the deep roots of legume crops penetrate far into the soil and use nutrients and moisture from deeper soil layers. Consequently, a higher total nitrogen uptake per unit of land in intercrops is obtained, compared to sole crops (Lithourgidis et al., 2010). It has been also reported that, after the harvest of legume crops, a considerable amount of residual nitrogen for consequent crops

Table 2.9 Soil available nitrogen and percentage of organic matter after harvest of wheat cropping systems under the studied cultivation methods in both growing seasons

	Soil available nitrogen (ppm)			Organic matter (%)		
	Broadcast in basins	Furrows	Raised beds	Broadcast in basins	Furrows	Raised beds
Sole	20.00	20.80	21.10	1.50	1.51	1.53
FIW1	21.30	21.91	22.42	1.55	1.56	1.59
FIW2	22.50	22.90	23.21	1.61	1.62	1.65
FIW3	23.80	24.04	24.40	1.66	1.67	1.70

FIW1 = 15% fahl clover intercropped with 100% wheat, FIW2 = 25% fahl clover intercropped with 100% wheat, FIW3 = 35% fahl clover intercropped with 100% wheat

was found, which provides some amount of organic matter enriching soil fertility (Banik et al., 2018).

The results in Table 2.9 revealed that soil available nitrogen was increased as planting density of fahl clover increased. The highest value of soil available nitrogen was obtained when 35% of fahl clover was intercropped with wheat on raised beds, namely 24.40 ppm. Whereas, the lowest value of soil available nitrogen was obtained when sole wheat broadcasted in basins, namely 20.00 ppm. Similarly, the highest value of organic matter percentage was obtained when 35% of fahl clover was intercropped with wheat on raised beds, namely 1.70%. Whereas, the lowest value of organic matter percentage was obtained when sole wheat broadcasted in basins, namely 1.50%.

Conclusion

To sustainably increase wheat yield, increasing land and water use efficiency should be implemented. This goal can be realized by breeding for resilient wheat cultivars can tolerate biotic and abiotic stresses, as well as employing improved management practices that can efficiently make use of the available resources to increase wheat yield. Inclusion of legume crops within the wheat production system can improve its yield, soil fertility and rationalize water use. The results showed that intercropping fahl clover with wheat on raised beds increase its yield by 17% and contributed in up to 30% saving in the applied irrigation water. It also increased soil available nitrogen and soil organic matter content by 7 and 1%, respectively. Thus, inclusion of legume crop in the production system of wheat by intercropping could attain the sustainable use of land by increasing soil available nitrogen and organic matter. It attained the sustainable use of water by using the applied water to wheat to produce fahl clover and wheat. Furthermore, it increased land productivity through increasing wheat yield, compared to wheat sole cultivation and producing two crops from the same piece of land.

References

Abbady, K. A., El-Maaz, E. I. M., Ahmed, H. M. R. M., & Zohry, A. A. (2016). Carbon sequestration as a function of intercropping management practices and different nitrogenous fertilizer types. *Journal of Soil Science and Agriculture Engineering, 7*(8), 565–586.

Abd El-Samie, F. S., Rady, M. O. A., Mahdi, A. H. A., & El-Hassan, M. E. (2018). Improving wheat productivity by application types different of fertilizer under newly reclaimed soil conditions. In *Egyptian Journal of Agronomy. The 15th international conference crop science*, pp. 31–40.

Abdelghany, A. M., Abouzied, H. M., & Badran, M. S. (2016). Evaluation of some Egyptian wheat cultivars under water stress condition in the north western coast of Egypt. *Journal of Agricultural and Environmental Sciences, 15*(1), 63–84.

Abdelmageed, K., Chang, X., Wang, D., Wang, Y., Yang, Y., Zhao, G., & Tao, Z. (2018). Evolution of varieties and development of production technology in Egyptian wheat: A review. *Journal of Integrative Agriculture 2019, 18*(3), 483–495.

Abdelrhim, A., Abd-Alla, H. M., Abdou, E., Ismail, M. E., & Cowger, C. (2018). Virulence of Egyptian blumeria graminis f. sp. tritici population and response of Egyptian wheat cultivars. *Plant Disease, 102*(2), 391–397.

Abd-Rabboh, A. M. K., & Koriem, M. H. M. (2021). Increasing unit area productivity and net income by intercropping peas with some wheat cultivars under different levels of nitrogen fertilizer. *Journal of the Advances in Agricultural Researches, 27*(2), 260–275.

Abou, G. A., & Tabl, K. M. (2014). Heat tolerance in some Egyptian wheat cultivars as measured by membrane thermal stability and proline content A. *Middle East Journal of Agriculture Research, 3*(2), 186–193.

Abouelenein, R., Oweis, T., El Sherif, M., Awad, H., Foaad, F., Abd El Hafez, S., Hammam, A., Karajeh, F., Karo, M., & Linda, A. (2009). Improving wheat Water productivity under different methods of irrigation management and nitrogen fertilizer rates. *Egyptian Journal of Applied Science., 24*(12A), 417–431.

Abouelenein, R., Oweis, T., El Sherif, M., Awad, H., Khalil, F.A., Abd El-Hafez, S., & Karajeh, F. (2010). A new water saving and yield increase method for growing berseem on raised seed bed in Egypt. *Egypt J. App. Sci. 25*(2A), 26–41.

Abou-Keriasha, M. A., Eisa, N. M. A., & Lamlom, M. M. (2013). Benefits of legume crops in rotation and intercropping for increased production and land use. *Egyptian Journal of Agronomy, 35*(2), 183–197.

Abou-Kerisha, M. A., Gadallah, R. A., & Badr, M. M. A. (2008). Effect of preceding and intercropping crops on yield and yield components of wheat. *Minufiya Journal of Agricultural Research, 33*(3), 709–728.

Aegic. (2015). Global grain market series: Egypt. Australian Export Grain Innovation Center. April 2015.

Agamy, R. A., Mohamed, G. F., & Rady, M. M. (2012). Influence of the application of fertilizer type on growth, yield, anatomical structure and some chemical components of wheat (Triticum aestivum L.) grown in newly reclaimed soil. *Australian Journal of Basic and Applied Sciences, 6*(3), 561–570.

Aggarwal, P., & Goswami, B. (2003). Bed planting system for increasing water use efficiency of wheat (T. Aestibum) grown in Inseptisol. *Indian Journal of Agricultural Sciences, 73*, 422–425.

Ahmad, I. M., Qubal, B., Ahmad, G., & Shah, N. H. (2009). Maize yield, plant tissue and residual soil N as affected by nitrogen management and tillage system. *Journal of Agricultural and Biological Sciences, 1*(1), 19–29.

Ali, A. M. (2018). *Effect of intercropping Egyptian clover variety fahl with wheat on yield and its components under some agricultural treatments* [PhD thesis. Assuit University. Egypt].

Ali, A. M., Shalaby, E. M., Ali, E. A., & Ibrahim, S. T. (2017). Impact of intercropping Egyptian clover (Trifolium Alexandrinum var fahl) with wheat on water use efficiency of irrigation. *Journal of Applied Life Sciences International, 14*(2), 1–12.

Allen, R. G., Pereira, L. S., Raes, D., & Smith, M. (1998). Crop evapotranspiration: Guideline for computing crop water requirements. FAO No. 56, pp. 109–158.

Andersen, M. K. (2005). *Competition and complementarily in annual intercrops—the role of plant available nutrients* [Ph.D. Thesis, Department of Soil Science, Royal Veterinary and Agricultural University, Copenhagen, Denmark. Samfunds litteraur Grafik, Frederiksberg, Copenhagen].

Atia, M. A. M., El-Khateeb, E. A., Abd El-Maksoud, R. M., Abou-Zeid, M. A., Salah̃, A., & Abdel-Hamid, A. M. E. (2021). Mining of leaf rust resistance genes content in Egyptian bread wheat collection. *Plants, 10*(7), 1378. https://doi.org/10.3390/plants10071378

Attia, M. A., & Abd El Salam, A. A. (2016). Effect of mineral, organic and bio-fertilizer on yield and yield components of bread wheat at Siwa Oasis Alex. *Journal of Agricultural Science, 61*(3), 211–219.

Awad, M. M., El-Mehy, A., & Hassan, H. H. (2023). Impact of crop sequence and nitrogen fertilization levels on wheat (Triticum aestivum L.) productivity in Egypt. *Egyptian Journal of Agricultural Research, 101*(1), 83–94.

Bahnas, O. T., & Bondok, M. Y. (2010). Effect of precision land leveling on faba bean response to compost application in sandy soils. *Misr Journal of Agricultural Engineering, 27*(2), 465–481.

Baka, Z. A. M. (2014). Plant extract control of the fungi associated with different Egyptian wheat cultivars grains. *Journal of Plant Protection Research, 54*(3), 231–237.

Bakheit, B. R., Abo-Elwafa, A., Abdel-Galil, M. M., & Abdelmonem, A. M. A. (2016). Impacts of recurrent selection and synthetic population on forage and seed yields of monocut Egyptian Clover (Trifolium alexandrinum L.). *Assiut Journal of Agricultural Science, 47*(1), 31–44.

Banik, D., Upadhyay, H., Devi, B. P., & Darvhankar, M. (2018). Effect of different irrigation system on growth and development of Wheat–Chickpea intercropping. *Bulletin of Environmental, Pharmacology and Life Sciences, 8*(1), 32–34.

Bertholdsson, N. O. (2005). Early vigor and Allelopathy—Two useful traits for enhancing barley and wheat competitiveness against weeds. *Weed Research, 45*, 94–102.

Bezabeh, M. W., Haile, M., Sogn, T. A., & Eich-Greatorex, S. (2022). Wheat (Triticum aestivum) production and grain quality resulting from compost application and rotation with faba bean. *Journal of Agriculture and Food Research, 10*, 100425.

CAPMAS. (2022). Central agency of public mobilization and statistics. Annual statistical year book 2022, Egypt.

Cong, W. F., Hoffland, E., Li, L., Six, J., Sun, J. H., & Bao, X. G. (2015). Intercropping enhances soil carbon and nitrogen. *Global Change Biology, 21*, 1715–1726. https://doi.org/10.1111/gcb.12738

Dababat, A. A., Muminjanov, H., & Smiley, R. W. (2015). Nematodes of small grain cereals: Current status and research. FAO.

Delgado, J. A., Nearing, M. A., & Rice, C. W. (2013). Conservation practices for climate change adaptation. *Advances in Agronomy, 121*, 47–115.

Dey, D., Nath, D., & Jamatia, P. B. (2015). Effect of raised bed planting method of maize under sandy loam soil of West Tripura. *International Journal of Applied Research, 1*(7), 561–563.

Dogan, E., & Kirnak, H. (2010). Water temperature and system pressure effect on drip lateral properties. *Irrigation Science*, 407–419.

Donn, S., Wheatley, R. E., McKenzie, B. M., Loades, K. W., & Hallett, P. D. (2014). Improved soil fertility from compost amendment increases root growth and reinforcement of surface soil on slope. *Ecological Engineering, 71*, 458–465.

Draz, I. S., Darwish, A. K., Abou-Elseoud, M. S., Elassal, A. A., & Komeil, D. A. (2021). Molecular discovery of new allele associated with loose smut resistance gene Ut-X in spring wheat. *Agronomy Research, 19*(1), 74–82. https://doi.org/10.15159/AR.21.055

Dwivedi, S. L., Sahrawat, K. L., Upadhyaya, H. D., et al. (2015). *Advances in host plant and rhizobium genomics to enhance symbiotic nitrogen fixation in grain legumes*. Elsevier Ltd.

El-Dessouki, W. A., Mohsena, R. K., Mansour & Eryan, N. L. (2022). Effects of certain weather, biotic factors and chemical components on the population of aphids in Egyptian wheat fields. *Egyptian Academy Journal of Biological Science, 15*(1), 1–13.

El-Hadidi, E. M., Ibrahim, M. M., Abdel-hafez, S. A., & Eid, M. S. M. (2014). Effect of deficit irrigation and raised bed on wheat yield, water productivity and water saving in North Nile Delta, Egypt. *Journal of Soil Science and Agricultural Engineering, Mansoura University, 6*(7), 845–862

El-Shamy M. A., EL-Mansoury, M. A. M., & Kasem, A. A. (2023). Effect of different sowing dates under intercropping patterns for fahl berseem and wheat crop under the north middle Nile Delta region. *Middle East Journal of Applied Sciences, 13*, 130–142.

El-Shamy, M. M., Sallam, M. E. A., & Awad, H. M. F. (2012). Powdery mildew infection on some Egyptian bread wheat cultivars in relation to environmental conditions. *Journal of Plant Protection and Pathology, 3*, 363–372.

Emam, M. A., Abd EL-Mageed, A. M., Niedbała, G., Sabrey, S. A., Fouad, A. S., Kapiel, T., Piekutowska, M., & Mahmoud, S. A. (2022). Genetic characterization and agronomic evaluation of drought tolerance in ten Egyptian wheat (Triticum aestivum L.) cultivars. *Agronomy, 12*, 1217. https://doi.org/10.3390/agronomy12051217

Esmail, S. M., Draz, I. S., Ashmawy, M. A., & El-Orabey, W. M. (2021). Emergence of new aggressive races of Puccinia striiformis f. sp. tritici causing yellow rust epiphytotic in Egypt. *Physiological and Molecular Plant Pathology, 114*, 101612. https://doi.org/10.1016/j.pmpp.2021.101612

Farag, F., Ghebrial, E., Abou Zeid, M., & Atwa, A. (2022). First report of Stemphylium leaf spot of wheat caused by Stemphylium vesicarium in Egypt. *Egyptian Journal of Phytopathology, 50*(1), 20–20. https://doi.org/10.21608/ejp.2022.125608.1056

Farhad, W., Saleem, M. F., Cheema, M. A., & Hammad, H. M. (2009) Effect of poultry manure levels on the productivity of spring maize (Zea mays L.). *The Journal of Animal and Plant Sciences, 19*(3), 122–125.

Gadallah A. M., Milad, I., Mabrook, S., & Abo Yossef, Y. M. (2017). Evaluation of some Egyptian bread wheat (triticum aestivum) cultivars under salinity stress. *Alexandria Science Exchange Journal, 38*, 259–270. https://doi.org/10.21608/asejaiqjsae.2017.3471

Gebrel, E. M. A., Gad, M. A., & Al-Farouk, M. O. (2019). Response of some wheat cultivars to different nitrogen fertilizer rates and their relation to rust diseases. *Egyptian Journal of Agronomy, 41*(3), 243–254.

Grant, R. F., Kimball, B. A., Conley, M. M., White, J. W., Wall, G. W., & Ottman, M. J. (2011). Controlled warming effects on wheat growth and yield: Field measurements and modeling. *Agronomy Journal, 103*(6), 1742–1754. https://doi.org/10.2134/agronj2011.0158

Guinet, M., Nicolardot, B., & Anne-Sophie, V. (2020). Nitrogen benefits of ten legume pre-crops for wheat assessed by field measurements and modelling. *European Journal of Agronomy, 120*, 126151.

Hafez, Y., Mazrou, Y., Shahin, A., Mehiar, F., Eid, M., & Abdelaal, K. (2022). Yield losses in wheat genotypes caused by stripe rust (Puccinia striifarmis f. sp. tritici) in North Delta, Egypt. *Notulae Botanicae Horti Agrobotanici Cluj-Napoca, 50*(2), 12622. https://doi.org/10.15835/nbha50212622

Hamada, A., & Hamd-Alla, W. (2019). Productivity of intercropped wheat with faba bean under crop sequences and foliar application of humic acid. *Egyptian Journal of Agronomy, 41*(3), 225–241. https://doi.org/10.21608/agro.2019.14396.1167

Hamada, A., Said, M. T., Ibrahim, K. M., Saber, M., & Sayed, M. A. (2022). A predictive study of the redistribution of some bread wheat genotypes in response to climate change in Egypt. *Agronomy, 12*, 113. https://doi.org/10.3390/agronomy12010113

Hamdany, M. Kh., Sheha, A. M., & El–Kalawy, S. M. S. (2016). Effect of crop sequence of rice, maize and fahll berseem residues on wheat productivity and soil fertility plant production. *Journal of Plant, 7*(12), 1401–1410.

Heiba, S. A. A., Osman, S. A., Eldessouky, S. E. I., et al. (2021). Genetic and biochemical studies on some Egyptian wheat genotypes under drought stress. *Bulletin of the National Research Centre, 45*, 151. https://doi.org/10.1186/s42269-021-00610-1

Helmi, A., & Rashwan, R. (2013). Effect of wheat cultivars and sown dates on aphid infestation in Egypt. *Munis Entomology & Zoology, 8*(2), 825–830.

Hesse, P. R. (1971). *A text book of soil chemical analysis*. Juan Murry (Publisher) Ltd.

Hossam El-Din, A. T. S. (2007). *Productivity of some wheat varieties by using bio and organic fertilization in the New Valley* [M.Sc. Thesis, Ain Shams University, Egypt].

Ibrahim, M., Ullah, H., Ahmad, B., & Malik, M. F. A. (2012). Effect of incremental dose of phosphorous and sulphur upon yield and protein content of wheat. *Biological Diversity and Conservation, 5*(3), 76–81.

Israelsen, O. W., & Hansen, V. E. (1962). *Irrigation principles and practices* (3rd ed.). Wiley International Edition.

Ivanov, S., Fedorova, E. E., Limpens Production, E., De Mita, S., Genre, A., Bonfante, P., et al. (2012). Rhizobium–legume symbiosis shares an exocytotic pathway required for arbuscule formation. *Proceedings of the National Academy of Sciences, 109*, 8316–8321. https://doi.org/10.1073/pnas.1200407109

Kandil, E. E., Schulz, R., & Müller, T. (2013). Response of some wheat cultivars to salinity and water stress. *Journal of Applied Sciences Research, 9*(8), 4589–4596.

Karrou, M., Oweis, T., Abou El Enein, R., & Sherif, M. (2012). Yield and water productivity of maize and wheat under deficit and raised bed irrigation practices in Egypt. *African Journal of Agricultural Research, 7*(11), 1755–1760.

Katare, S., Singh, B., Patil, S. D., Tiwari, R., Jasrotia, P., Saharan, M. S., & Sharma, I. (2015). Evaluation of new insecticides for management of foliar aphid complex in wheat. *Indian Journal of Entomology, 79*(2), 185–190.

Khalil, F. A. F., & Abouenein, R. (2012). Rationalization of irrigation water in agricultural sector through integrated water management and community participation. *Fayoum Journal of Agricultural Research and Development, 26*(2), 11–20.

Kirihetti, D. (2018). *Direct assessment of the fixation and release of nitrogen by pea, lentil, chickpea and faba bean* [A Master of Science Thesis, University of Saskatchewan Saskatoon, Canada].

Klute, A. (1986). Methods of soil analysis, Part I. Physical and mineralogical method. American Society of Agronomy.

Leng, G., & Hall, J. (2019). Crop yield sensitivity of global major agricultural countries to droughts and the projected changes in the future. *Science of the Total Environment, 654*, 811–821.

Li, C. J., Hoffland, E., Kuyper, T. W., Yu, Y., Li, H., Zhang, C., Zhang, F., & van der Werf, W. (2020). Yield gain, complementarity and competitive dominance in intercropping in China: A meta-analysis of drivers of yield gain using additive partitioning. *European Journal of Agronomy, 113*, 125987. https://doi.org/10.1016/j.eja.2019.125987

Li, Y., Zhang, W., Ma, L., Wu, L., Shen, J., Davies, W. J., Oenema, O., Zhang, F., & Dou, Z. (2014). An analysis of China's grain production: Looking back and looking forward. *Food and Energy Security, 3*, 19–32.

Lithourgidis, A. S., & Dordas, C. A. (2010). Forage yield, growth rate, and nitrogen uptake of faba bean intercrops with wheat, barley, and rye in three seeding ratios. *Crop Science, 50*, 2148–2158. https://doi.org/10.2135/cropsci2009.12.0735

Mabrouk, O. I., Fahim, M. A., Abd El Badeea, O. E., & Omara, R. I. (2022). The impact of wheat yellow rust on quantitative and qualitative grain yield losses under Egyptian field conditions. *Egyptian Journal of Phytopathology, 50*(1), 1–19. https://doi.org/10.21608/ejp.2022.117996.1054

Majeed, A., Muhmood, A., Niaz, A., Javid, S., Ahmad, Z. A., Shah, S. S. H., & Shah, A. H. (2015). Bed planting of wheat (Triticum aestivum L.) improves nitrogen use efficiency and grain yield compared to flat planting. *The Crop Journal, 3*, 118–124.

Majumder, D., Rajesh, T., Suting, E. G., & Debbarma, A. (2013). Detection of seed borne pathogens in wheat: Recent trends. *Australian Journal of Crop Science, 7*(4), 500–507.

Mansour, S. (2012). Global agriculture information system. Egypt: Wheat and corn production on the rise: Grain and feed annual. Annual report.

Mao, L. L., Zhang, L. Z., Li, W. W., Werf, W. V. D., Sun, J. H., Spiertz, H., & Li, L. (2012). Yield advantage and water saving in maize/pea intercrop. *Field Crops Research, 138*, 11–20.

Mat Hassan, H., Marschner, P., McNeill, A., & Tang, C. (2012). Grain legume pre-crops and their residues affect the growth, P uptake and size of P pools in the rhizosphere of the following wheat. *Biology and Fertility of Soils, 48*, 775–785. https://doi.org/10.1007/s00374-012-0671-8

Mead, R., & Willey, R. W. (1980). The concept of a "land equivalent ratio" and advantages in yields from intercropping. *Experimental Agriculture, 16*, 217–228.

Megahed, M. M. M., & Kamel, H. M. (2020). Susceptibility of certain Egyptian wheat cultivars to greenbug, Schizaphis graminum (Rondani) (Homoptera: Aphididae) Infestation. *Egyptian Academy Journal of Biological Sciences, 13*(4), 247–257.

Melad, N. M. K. (2018). *Evaluation of some crop sequences of rice on wheat productivity* [MSc thesis, Kafer EL-Sheikh University, Egypt].

Mohamed, M. F., Thalooth, A. T., Elewa, T. A., et al. (2019). Yield and nutrient status of wheat plants (Triticum aestivum) as affected by sludge, compost, and biofertilizers under newly reclaimed soil. *Bulletin of the National Research Centre, 43*, 31. https://doi.org/10.1186/s42269-019-0069-y

Morris, R. A., & Garrity, D. P. (1993). Resource capture and utilization in intercropping: Water. *Field Crops Research, 34*, 303–317.

Nassar, R., Kamel, H. A., Ghoniem, A. E., Alarcón, J. J., Sekara, A., Ulrichs, C., et al. (2020). Physiological and anatomical mechanisms in wheat to cope with salt stress induced by seawater. *Plants, 9*, 237. https://doi.org/10.3390/plants9020237

Omara, R. I., El-Naggar, D. R., Abd El-Malik, N. I., & Ketta, H. A. (2016). Losses assessment in some Egyptian wheat cultivars caused by stripe rust pathogen (Puccinia striiformis). *Egyptian Journal of Phytopathology, 44*(1), 199–203.

Ouda S., & Zohry, A. A. (2017). Crops intensification to reduce wheat gap in Egypt. In *Future food gaps in Egypt: Obstacles and opportunities*. Springer Publishing House. ISBN 978-3-319-46942-3.

Ouda, S., Zohry, A., & Abdel-Wahab, S. (2020). Sustainable intensive cropping to reduce irrigation-induced erosion: II. Changing cropping sequence under sprinkler irrigation practice. *Moroccan Journal of Agricultural Science, 1*(1), 26–33.

Pane, C., Celano, G., Piccolo, A., Villecco, D., Spaccini, R., Palese, M., & Zaccardelli, M. (2015). Effects of on-farm composted tomato residues on soil biological activity and yields in a tomato cropping system. *Chemical and Biological Technologies in Agriculture, 2*, 4.

Porto, V. C. N., Bezerra, N. F., Lima, J. S. S., Barros, J. A. P., & Moreira, J. N. (2011). Combination of lettuce and rocket cultivars in two cultures intercropped with carrots. *Horticultura Brasileira, 29*, 404–411.

Preissel, S., Reckling, M., Schläfke, N., & Zander, P. (2015). Magnitude and farm-economic value of grain legume pre-crop benefits in Europe: A review. *Field Crops Research, 175*, 64–79. https://doi.org/10.1016/j.fcr.2015.01.012

Ragab, Kh. E., & Kheir, A. M. S. (2019). Characterizing some Egyptian bread wheat cultivars for salinity tolerance. *Journal of Plant Production, 10*(12), 1043–1049. https://doi.org/10.21608/jpp.2019.71545

Rao, M. R., & Willey, R. W. (1980). Evaluation of yield stability in intercropping: Studies on sorghum/pigeonpea. *Experimental Agriculture, 16*(2), 105–116.

Rekaby, S. A., Eissa, M. A., Hegab, S. A., & Ragheb, H. M. (2016). Effect of nitrogen fertilization rates on wheat grown under drip irrigation system. *Assiut Journal of Agricultural Science, 47*(3), 104–119.

Rücknagel, J., Götze, P., Koblenz, B., Bachmann, N., Löbner, S., Lindner, S., Bischoff, J., & Christen, O. (2016). Impact on soil physical properties of using large-grain legumes for catch crop cultivation under different tillage conditions. *European Journal of Agronomy, 77*, 28–37.

Scotti, R., Ascoli, D. R., Bonanomi, G., Caceres, M. G., Sultana, S., Cozzolino, L., Scelza, R., Zoina, A., & Rao, M. A. (2015). Combined use of compost and wood scraps to increase carbon stock and improve soil quality in intensive farming systems. *European Journal of Soil Science*. https://doi.org/10.1111/ejss.12248

Sheha, A. M., Abdel-Wahab, T. I., & Abdel-Wahab, S. I. (2015). Maximizing nitrogen and land use efficiencies of intercropped wheat with pea under different pea sowing dates. *Journal of Plant Sciences, 3*(6), 358–371. https://doi.org/10.11648/j.jps.20150306.20

Shen, H., Zhong, X., Zhao, F., Wang, Y., Yan, B., Li, Q., Chen, G., Mao, B., Wang, J., Li, Y., Xiao, G., He, Y., Xiao, H., Li, J., & He, Z. (2015). Overexpression of receptor-like kinase ERECTA improves thermotolerance in rice and tomato. *Nature Biotechnology, 33*(9), 996–1003.

Shenoda, J. E., Sanad, M. N. M. E., Rizkalla, A. A., El-Assal, S., Ali, R. T., & Hussein, M. H. (2021). Effect of long-term heat stress on grain yield, pollen grain viability and germinability in bread wheat (Triticum aestivum L.) under field conditions. *Hyliyon, 7*(6), e07096.

Sing, V. K., Dwivedi, B. S., Shukla, A. K., & Mishra, R. P. (2010). Permanent raised bed planting of the pigeon pea-wheat system on a Typic Ustoochrept: Effects on soil fertility, yield and water and nutrient use efficiencies. *Field Crops Research, 116*, 127–139.

Sorour, S. G. H., Aiad, M. A., Ahmed, A. A., Henash, M. I. A., Metwaly, E. M., Alharby, H., et al. (2019). A yield of wheat is increased through improving the chemical properties, nutrient availability and water productivity of salt affected soils in the North Delta of Egypt. *Applied Ecology and Environmental Research, 17*(4), 8291–8306.

Swelam, A. (2016). Raised-bed planting in Egypt: An affordable technology to rationalize water use and enhance water productivity. https://repo.mel.cgiar.org/

Tan, K. H. (1996). *Soil sampling preparation and analysis* (p. 391). Marcel Dekker, Inc.

Tembo, B., Mulenga, R. M., Sichilima, S., Msiska, K. K., Mwale, M., & Chikoti, P. C. (2020). Detection and characterization of fungus (Magnaportheoryzae pathotype Triticum) causing wheat blast disease on rain-fed grown wheat (Triticum aestivum L.) in Zambia. *PlosOne, 15*(9), e0238724.

Thabet, M., & Najeeb, K. M. A. (2017). Impact of wheat leaf rust severity on grain yield losses in relation to host resistance for some Egyptian wheat cultivars. *Middle East Journal of Agriculture Research, 06*(04), 1501–1509.

USDA. (2022). Decline in Ukraine wheat imports drives Egypt to diversify its suppliers. United State Department of Agriculture. Report Number: EG2022-0015.

Vermeiren, L., & Jopling, G. A. (1984). Localized irrigation. FAO, Irrigation and Drainage Paper no. 36, Rome, Italy.

Walkley, A., & Black, I. A. (1934). An examination of Degtjareff method for determining organic carbon in Soils: Effect of variations in digestion conditions and of inorganic soil constituents. *Soil Science, 63*, 251–263.

Wang, Y. F., Qin, Y. Z., Chai, Q., Feng, F. X., Zhao, C., & Yu, A. Z. (2018). Interspecies interactions in relation to root distribution across the rooting profile in wheat-maize intercropping under different plant densities. *Frontiers in Plant Science*, 9483–9489.

Wick, A., Berti, M., Lawley, Y., & Liebig, M. (2017). Integration of annual and perennial cover crops for improving soil health. In *Soil health and intensification of agroecosystems* (pp. 127–150). Elsevier BV.

Yang, X., Ni, K., Shi, Y., Yi, X., Zhang, Q., Fang, L., Ma, L., & Ruan, J. (2018). Effects of long-term nitrogen application on soil acidification and solution chemistry of a tea plantation in China. *Agriculture, Ecosystems & Environment, 252*, 74–82.

Yin, W., Fan, Z. L., Hu, F. L., Fan, H., Yu, A. Z., Zhao, C., Chai, Q. (2019). Straw and plastic mulching enhances crop productivity via optimizing interspecific interactions of wheat–maize intercropping in arid areas. *Crop Science, 59*, 2201–2213.

Zakaria, M. M. (2020). Response of some bread wheat genotypes to heat stress. *Egyptian Journal of Plant Breeding, 24*(1), 137–158.

Zhang, P., Liu, Y., Liu, W., Cao, M., Massart, S., & Wang, X. (2017). Identification, characterization and full-length sequence analysis of a novel polerovirus associated with wheat leaf yellowing disease. *Frontiers in Microbiology, 8*, 1689.

Zohry, A. A., Ouda, S., Hamd-Alla, W., & Shalaby, E. (2017a). Evaluation of different crop sequences for wheat and maize in sandy soil. *Acta Agriculturae Slovenica., 109*(2), 383–392.

Zohry, A., Abbady, K., El-Mazz, E., & Ahmed, H. (2017b). Maximizing land productivity by diversified cropping systems with different nitrogen types. *Acta Agriculturae Slovenica, 109*, 482–492.

Zohry, A., Ouda, S., & Abdel-Wahab, T. (2020). Sustainable intensive cropping to reduce irrigation-induced erosion: I. Intercropping systems under surface irrigation practice. *Moroccan Journal of Agricultural Science, 1*(2), 63–71.

Zohry, A., & Ouda, S. (2019). Intercropping systems for sugar beet to improve its land and water productivity. *Journal of Soil Crops, 29*(2), 218–226.

Zohry A., & Ouda S. (2020). Wheat insufficiency and deficit irrigation. In *Deficit irrigation: A remedy for water scarcity*. Springer Publishing House. ISBN 978-3-030-35585-2.

Chapter 3
Climate Variability and Change Disturbs Maize Production

Abstract Climate variability is expected to cause instability in the yield of the cultivated crops, which creating a high risk to global crops production. In Egypt, maize is one of the main cereal crops used as human food, where its flour mixed with wheat flour by 20% to produce bread. The breeding programs in Egypt were successful to produce high yielding and disease resistance maize hybrids possess yield stability traits. In this chapter, we analyzed the relationship between some weather elements (solar radiation, maximum and minimum temperatures, and relative humidity) and maize productivity in the five agro-climatic zones of Egypt from 2009 to 2021 using multiple linear regression analysis. The analysis of the studies weather elements and zonal- specific maize productivity revealed that maize is most suitable to grow in the fourth agro-climatic zone, because there was less fluctuation in its productivity there, compared to its value in the rest of the agro-climatic zones. The regression analysis between maize productivity and the studied weather elements suggested that the main seasonal weather elements affecting maize productivity in the five agro-climatic zones of Egypt are seasonal solar radiation in the first, second and fifth agro-climatic zones. The seasonal maximum temperature is also an affecting factor in maize productivity in the first, second and fourth agro-climatic zones. The seasonal minimum temperature was found to have an effect on maize productivity in the first, third and fifth agro-climatic zones. Seasonal relative humidity was found to be effective in the third agro-climatic zone only. Furthermore, the projected effect of climate change on maize productivity was assessed in the five agro-climatic zones of Egypt using CMIP6 mean projection of two future scenarios, namely SSP1-2.6 and SSP5-8.5 in two time-intervals (2060–2079 and 2080–2099), where maximum and minimum temperatures, as well as relative humidity were downloaded. Both scenarios projected a gradual increase in the maximum and minimum temperature from the first to the fifth agro-climatic zones, whereas seasonal relative humidity will be gradually decrease toward the fifth agro-climatic zone. The regression of the studied weather elements on maize productivity in the time period from 2009 to 2021 resulted in developing a zonal-specific prediction equations to be used in the projection of maize productivity in the future under the two scenarios. The results showed that maize productivity will be gradually reduced under both future climate scenarios, where the lowest reduction will occur in the first agro-climatic zone and the highest

S. Ouda and A. E. Zohry, *Integration of Legume Crops with Cereal Crops Under Changing Climate*, https://doi.org/10.1007/978-3-031-68102-8_3

reduction will occur in the fifth agro-climatic zone. Additionally, a brief review was presented on the effect intercropping legume crops with maize on relieving the consequences of climate change on maize productivity.

Keywords Agro-climatic zones of Egypt · CMIP6 · SSP1-2.6 · SSP5-8.5 · Zonal-specific prediction equations · Maize intercropping systems with legume crops

Introduction

Maize is an important cereal crop ranks the third after wheat and rice. It is recognized as an important human food crop in many countries worldwide, where it contributes over 20% of food calories in these countries (Shiferaw et al., 2011). Maize is a multi-purpose crop, where it also used as a feed crop and most recently as an energy crop (Poole et al., 2021). Maize global cultivated area was increased 46%, its production was increased by 118% and productivity was increased by 50% between 1995 and 2019 (Erenstein et al., 2022).

Climate variability is expected to cause instability in crops yields, which creating a high risk to global crop production (Lobell et al., 2013). Crop yields show a strong correlation with temperature change and with the duration of heat waves, which differ based on plant maturity stages during extreme weather events (Hoffmann, 2013). In spite of being a warm season crop, maize is more sensitive to high temperature stress than wheat (Tesfaye et al., 2015). Higher temperatures have negative impacts on maize yields with multiplication of some weeds and pests in the same time (Shiferaw et al., 2011). In maize adult plants, the increase in temperature slows down lateral root growth to promote the development of long axile roots to reach the water of the deeper soil layers (Hund et al., 2008). Acceleration in phenological development of maize under warmer conditions was observed, which resulted in reduction in the growing season, reduction in net photosynthesis and consequently reduction in productivity (Lobell et al., 2013). High temperature stress at critical developmental stages of maize plants causes significant yield loss (Lobell et al., 2011).

Maize plants become susceptible to high temperatures after reaching eight-leaf stage to seventeen-leaf stages (Chen et al., 2010), which has significant impact on plant growth, architecture, ear size and kernel numbers (Farré & Faci, 2006). Temperature above 35 °C for a prolonged period is considered unfavorable for maize crop growth and development and, 40 °C during flowering and grain filling have severe negative impacts on grain yield (Shiferaw et al., 2011). It has been reported that at a high temperature of 35 °C, maize yield reduces by 9% (Adhikari et al., 2015). Furthermore, higher air temperature than 36 °C causes pollen to lose viability in maize (Motha, 2011). A significantly reduction in stomatal conductance under heat stress was observed resulting in a reduced rate of photosynthesis, leaf area, reduced biomass accumulation and seed weight in maize (Zhao et al., 2016).

Considerable evidences have revealed that climate change is already affecting agricultural production systems worldwide. The Inter-Governmental Panel on Climate Change (IPCC) in its Fifth Assessment Report has warned that global climate has been changing and it would continue to happen in the near future, where the global mean surface temperatures are projected to increase by 1.4 to 5.8 °C by the end of this century relative to 1990 (IPCC, 2014). Such increase may have a significant negative influence on agricultural productivity (Bates et al., 2008), in addition to increasing drought, salinity, waterlogging, and mineral toxicity stresses (Teixeira et al., 2013). These changes will have a dramatic effect on agricultural systems especially in temperate countries, where the water shortage and high temperatures are the main limiting factors for the crop production (Rinaldi, 2009).

Climate change studies revealed that the change in the weather conditions will contribute in decline in maize production. Msowoya et al. (2016) stated that rain fed maize production in Malawi, may decrease to 14% by mid-century due to climate change, rising to as much as 33% loss by the end of the century. Zhao et al. (2017) projected reduction in maize yield by 7% for every degree increase in global mean temperature. Although new maize hybrids are developed by plant breeders adopted to biophysical environmental stresses, it might not be able to adopt to climate change stresses and its productivity still diminish (Banziger & Diallo, 2004).

In Egypt, maize is one of the main cereal crops used as human food, where its flour mixed with wheat flour by 20% to produce bread. It is used as feed for livestock and poultry by 70% of their fodder. Furthermore, it is used as row material for industrial products, such as oil and starch. The breeding programs in Egypt were successful to produce high yielding and disease resistance hybrids possess yield stability traits. There are two types of maize hybrids in Egypt, namely white and yellow grains hybrids. The Egyptian maize hybrids are suited to be grown throughout Egypt, in the clay soil of the old lands located on the banks of the Nile River, as well as in the sandy lands of the new reclaimed areas. Additionally, technical recommendations to attain high maize yield have been also provided to the farmers.

The local production of maize is insufficient to meet its excessive demand, especially on the yellow maize grains. This shortage is compensated by importing maize from other counties. Because expanding in the cultivated area with maize is time and money consuming, attempts have been made to improve management practices of maize to increase its national production.

Climate variability and change are negatively affecting maize national production. Saleh (2007) indicated that the national maize production was reduced by an average of 11% in 2015, as a result of heat wave occurred in that year, compared to its production in 2010. Furthermore, climate change studies in Egypt have been reported high decline in maize production. Some simulation studies on the field level revealed that the decline in maize production could be as high as 47% (Ouda et al., 2009), 46% (Moneir, 2012) and 38% (Taha, 2012), which depends on soil type and irrigation system. Furthermore, Morsy (2015) simulated the effect of climate change on maize yield using RCP6.0 scenario developed by CCSM4 model in 2030 and he projected 13–15% maize yield losses. The effect of climate change on maize national production was studied by Zohry and Ouda (2017), where they reported a decreased by 7% under climate change condition in 2030. Whereas, Gamal et al. (2021) projected a reduction in the national maize production by 4%.

In this chapter, we analyzed the relationship between some weather elements (solar radiation, maximum and temperatures, and relative humidity) and maize productivity in the five agro-climatic zones of Egypt using multiple linear regression analysis. This analysis will reveal the effect of climate variability on maize productivity in these agro-climatic zones. Furthermore, the projected effect of climate change on maize productivity was assessed in the five agro-climatic zones of Egypt using CMIP6 mean projection of two future scenarios, namely SSP1-2.6 and SSP5-8.5 in two time-intervals. Furthermore, a brief review was presented on the effect legume crops intercropping with maize on relieving the consequence of climate change on maize productivity.

Assessment of the Effect of Climate Variability on Maize Productivity

Collected Data

Ouda and Noreldin (2017) classified Egypt to five agro-climatic zones according to evapotranspiration values (ETo) for the agricultural governorates in Egypt located around the Nile River using a 10-year time period, where analysis of variance was employed and the means was separated and ranked using least significant difference test. Figure 3.1 presented the map of Egypt with its five agro-climatic zones.

Weather data for the five agro-climatic zones of Egypt were obtained from the following website: https://power.larc.nasa.gov. Daily weather data of solar radiation ($MJ/m^2/day$), maximum and minimum temperatures (°C), and relative humidity (%) from 2009 to 2021 was downloaded from the above website and monthly averages from the beginning of May to the end of August (maize growing season) were calculated. The seasonal values of the selected weather elements were calculated for each agro-climatic zone. The latitude, longitude and altitude of each agro-climatic zone are presented in Table 3.1.

In addition to the calculated seasonal weather elements, the corresponded average maize productivity values (ton/ha) in each zone were also collected from the data set of the Ministry of Agricultural and lands reclamation in Egypt. Hence, this analysis will determine the spatial-variability of zone-specific maize productivity from one agro-climatic zone to the other, as a result of the studied weather elements in each zone.

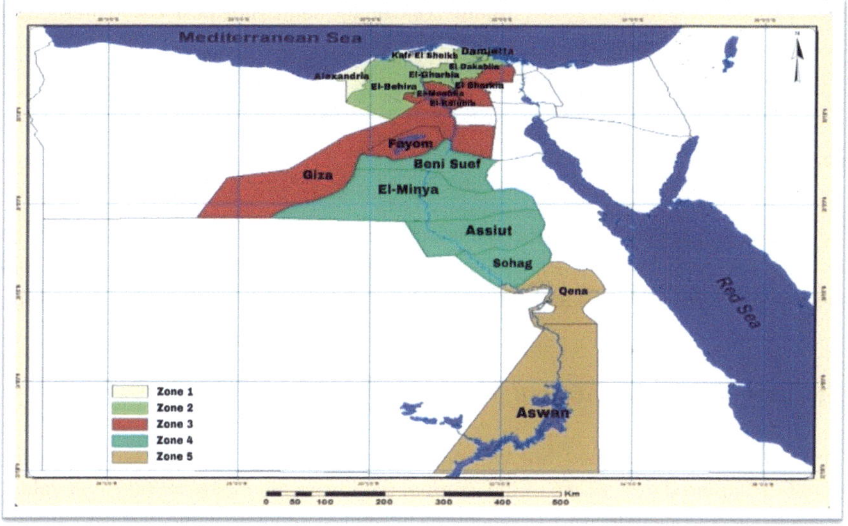

Fig. 3.1 Map of agro-climatic zones of Egypt. *Source* Ouda and Noreldin (2017)

Table 3.1 Latitude, longitude, altitude and ETo range of the agro-climatic zones of Egypt

Agro-climatic zones	Latitude (°)	Longitude (°)	Altitude (m)	ETo range
First agro-climatic zone	31.70–31.07	29.00–30.57	7.00–20.00	4.28–4.85
Second agro-climatic zone	31.25–30.47	30.28–32.14	5.00–14.80	5.12–5.34
Third agro-climatic zone	30.36–29.18	31.01–30.51	13.00–30.00	5.59–5.96
Fourth agro-climatic zone	29.04–26.36	30.44–31.38	30.40–71.00	6.12–6.14
Fifth agro-climatic zone	26.10–24.02	32.43–32.53	72.60–108.30	6.48–6.60

Source Ouda and Noreldin (2017)

Statistical Analysis

Descriptive statistics analysis was done and coefficient of variation (CV%) was calculated to assess the variability existed in the data set. CV (%) is a measure of relative variability. It is the ratio of the standard deviation to the mean and it shows the extent of variability in relation to the mean of the population. The higher the CV, the greater the dispersion (Gomez & Gomez, 1984).

Furthermore, multiple linear regression analysis between maize productivity as dependent variable and weather elements as independent variables was done using 95% confidence interval (Draper & Smith, 1987). The significant independent variables in the analysis were graphed to show its line fit between the measured and the predicted values of maize productivity. The multiple linear regression analysis was used to regress seasonal weather elements in the time period from 2009 to 2021 on maize productivity in each agro-climatic zone on the same time period and prediction equations were developed.

Goodness of Fit Test

The goodness of fit between the measured values of maize productivity and the predicted productivity values was tested by coefficient of determination (R^2), root mean square error (RMSE) and Willmott index of agreements (*d-stat*) as followed:

Coefficient of Determination (R^2)

The R^2 value is calculated from the total sum of squares, more precisely, it is the sum of the squared deviations of the original data from the mean. R^2 takes on values between 0 and 1.0, a value of 1.0 indicates a perfect match, and value of 0 indicates no agreement at all. R^2 also shows how many points fall on the regression line (Jamieson et al., 1998).

$$R^2 = 1 - \frac{\sum_{i=1}^{n}(O_i - S_i)^2}{\sum_{i=1}^{n}\left(O_i - \overline{O}\right)^2} \tag{3.1}$$

where O_i and S_i represent the observed and predicted values, respectively, n represents the number of observed and predicted values used in comparison and $\overline{O}$ is the observed average.

Root Means Square Error (RMSE)

It is the standard deviation of the residuals (prediction errors). Residuals are a measure of how far from the regression line data points are. It is a frequently used to measure the difference between values predicted by a model and those actually observed from an experiment that is being modeled (Jamieson et al., 1998).

$$\text{RMSE} = \sqrt{\left(\frac{\sum_{i=1}^{n}(S_i - O_i)^2}{n}\right)} \tag{3.2}$$

Willmott Index of Agreement (*d-stat*)

It is the standardized measure of the degree of model prediction error and varies between 0 and 1.0. A value of 1.0 indicates a perfect match, and value of 0 indicates no agreement at all (Willmott, 1981).

$$d\text{-}stat = 1 - \frac{\sum_{i=1}^{n}(O_i - S_i)^2}{\sum_{i=1}^{n}\left[\left(\left|(S_i - \overline{O})\right| + \left|(O_i - \overline{O})\right|\right)^2\right]} \tag{3.3}$$

Assessment of the Effect of Climate Change on Maize Productivity

The 6th phase of Coupled Model Intercomparison Project (CMIP6) is a new developed generation of global climate models (Eyring et al., 2016). One novelty of it is the Shared Socio-economic Pathways are combined with the CMIP5 global climate models (GCMs). These combinations aim to develop future climate change conditions by ensuring major improvements in emissions, land use scenarios, improved models for parameterization and physical process, among others (Eyring et al., 2016). The CMIP6 global climate models are more complex than previous models, with higher spatial resolution and improvements in physical process simulations (Kim et al., 2020).

The data of CMIP6 mean projection of two future scenarios, namely SSP1-2.6 and SSP5-8.5 were downloaded from The World Bank Group, Climate Change Knowledge Portal (CCKP) at the following website: https://climateknowledgeportal.worldbank.org/. The CCKP serves as a leading hub of global climate data, information, and tools to support decision making and improve the understanding of future climates and projected risks. The CCKP was designed to provide open access to climate data and related information in order to help users to better understand future climate scenarios and characterize potential risks and impacts with data aggregated at national, sub-national, and watershed levels (World Bank Group, 2021).

We collected projected yearly data of Egypt and sub-region from 2024 to 2049 for two future climate scenarios, namely SSP1-2.6 and SSP5-8.5 developed from a multi-model ensemble. Climate projection data is modeled data, derived from the Coupled Model Inter-comparison Projects. CMIP data is the foundational data used in the IPCC Assessment Reports, where CMIP6 supports the IPPC's Sixth Assessment Report, and initial release was in August 2021. Projection data is presented at a spatial resolution, $1^\circ \times 1^\circ$ (100 km × 100 km) (World Bank Group, 2021).

In CMIP6, future climate scenarios are presented through five SSPs: SSP1-1.9, SSP1-2.6, SSP2-4.5, SSP3-7.0, and SSP5-8.5, which present different societal development pathways. The total radiative forcing level by 2100 (the cumulative measure of GHG emissions from all sources) is presented at the end of each pathway (i.e., – 1.9, – 2.6, – 4.5, 7.0, 8.5, etc.). The description of the two selected future scenarios, namely SSP1-2.6 and SSP5-8.5 is as followed:

SSP1-2.6: supports increasing sustainability with global emissions cut severely, but reach net-zero after 2050.

SSP5-8.5: presents a future based on an intensified exploitation of fossil fuel resources, where global markets are increasingly integrated leading to innovations and technological progress (World Bank Group, 2021).

The downloaded data is divided to eight 10-year intervals starting in 2024 and ends 2094. We used this data to characterize the future climate of Egypt between these time intervals. Another data set was downloaded and used to project maize productivity under the two studied future climate scenarios. These data sets are composed of maximum and minimum temperatures and relative humidity as averages on four time-interval, namely 1995–2014, 2020–2039, 2040–2059, 2060–2079 and 2080–2099 and we used 1995–2014 as a baseline value to compare between the past values and the future projected values. We calculated the seasonal values maize growing season on agro-climatic zonal level in each time-interval. The downloaded data of the three climate elements from the two future climate scenarios was used to project maize productivity in the future using the developed prediction equations. The projected values of maize productivity were compared to a baseline value of measured maize productivity averaged from 2009 to 2021, which was obtained from the Ministry of Agricultural and Land Reclamation in Egypt.

Effect of Weather Variability on Maize Productivity in the Agro-climatic Zones of Egypt

The First Agro-climatic Zone

Descriptive statistics analysis for the studied seasonal weather elements and maize productivity in the first agro-climatic zone are presented Table 3.2. Regarding to weather elements, the results indicated that the highest value of the range between minimum and maximum values and the highest value of CV% were found for seasonal relative humidity, namely 8.13% and 5.30%, respectively. Whereas, the lowest value of the range between minimum and maximum values and the lowest value of CV% were found for seasonal minimum temperature, namely 3.15 °C and 4.24%, respectively. With respect to maize productivity, low value of the range between minimum and maximum values was found, namely 1.51 ton/ha and relatively high value of the CV%, namely 4.61%. These results implied that seasonal solar radiation, seasonal maximum and minimum temperature are less variable, and might have a significant effect on maize productivity in the first agro-climatic zone more than the effect of seasonal relative humidity.

Table 3.2 Descriptive statistics of seasonal weather elements and maize productivity (ton/ha) in the first agro-climatic zones of Egypt from 2009 to 2021

Statistic	SRAD	T_{max}	T_{min}	RH	Maize productivity
Minimum value	24.15	30.32	20.20	52.28	7.75
Maximum value	28.68	34.33	23.36	60.41	9.26
Range	4.52	4.01	3.15	8.13	1.51
Mean	27.16	32.04	21.91	56.45	8.68
Standard deviation	1.19	1.46	0.93	2.99	0.40
CV (%)	4.39	4.55	4.24	5.30	4.61

SRAD = solar radiation (MJ/m^2/day), T_{max} = maximum temperature (°C), T_{min} = minimum temperature (°C), RH = relative humidity (%)

The Second Agro-climatic Zone

The results in Table 3.3 revealed that the highest value of CV% was found for seasonal minimum temperature and seasonal relative humidity, namely 8.54 and 19.29%, respectively. Whereas, the lowest value of the range between minimum and maximum values and lowest CV% were found for seasonal solar radiation, namely 3.85 MJ/m^2/day and 4.42%, respectively. Regarding maize productivity, a low value of the range between minimum and maximum values was found, namely 1.25 ton/ha as well as relatively high value of the CV%, namely 4.89%. The above results suggest that seasonal solar radiation, and seasonal maximum temperature were less variable, and might have a significant effect on maize productivity in the second agro-climatic zone more than seasonal minimum temperature and seasonal relative humidity.

Table 3.3 Descriptive statistics of seasonal weather elements and maize productivity (ton/ha) in the second agro-climatic zones of Egypt from 2009 to 2021

Statistic	SRAD	T_{max}	T_{min}	RH	Maize productivity
Minimum value	24.44	31.30	18.86	38.86	7.78
Maximum value	28.29	37.87	25.50	63.01	9.03
Range	3.85	6.57	6.64	24.15	1.25
Mean	27.13	35.03	20.98	49.78	8.60
Standard deviation	1.20	2.16	1.79	9.60	0.42
CV (%)	4.42	6.16	8.54	19.29	4.89

SRAD = solar radiation (MJ/m^2/day), T_{max} = maximum temperature (°C), T_{min} = minimum temperature (°C), RH = relative humidity (%)

Table 3.4 Descriptive statistics of seasonal weather elements and maize productivity (ton/ha) in the third agro-climatic zones of Egypt from 2009 to 2021

Statistic	SRAD	T_{max}	T_{min}	RH	Maize productivity
Minimum value	24.98	36.70	19.61	34.03	7.44
Maximum value	27.94	40.71	22.06	39.99	9.10
Range	2.96	4.01	2.45	5.97	1.66
Mean	27.42	38.23	20.76	36.89	8.39
Standard deviation	0.77	1.33	0.69	1.69	0.54
CV (%)	2.83	3.49	3.34	4.58	6.38

SRAD = solar radiation (MJ/m^2/day), T_{max} = maximum temperature (°C), T_{min} = minimum temperature (°C), RH = relative humidity (%)

The Third Agro-climatic Zone

The results in Table 3.4 showed that less variability in the seasonal relative humidity in the third agro-climatic zone manifested by relatively low value of the range between minimum and maximum values and relatively low value of CV%, compared to its values in the first and second agro-climatic zones. The lowest value of CV% were found for seasonal solar radiation, followed by seasonal minimum temperature, namely 2.83 and 3.34%, respectively. Regarding maize productivity, low value of the range between minimum and maximum values, namely 1.66 ton/ha and relatively high value of the CV%, namely 6.38% were found. The above results suggest that seasonal solar radiation, and seasonal minimum temperature were less variable, and might have a significant effect on maize productivity in the third agro-climatic zone.

The Fourth Agro-climatic Zone

In the fourth agro-climatic zone (Table 3.5), the results revealed that the lowest value of the range between minimum and maximum values and CV% were found for seasonal solar radiation, namely 1.04 MJ/m^2/day and 0.91%, respectively. Whereas, the highest values of the range between minimum and maximum values and CV% were found for seasonal relative humidity, namely 6.12% and 7.01%, respectively. With respect to maize productivity, a low value of the range between minimum and maximum values, namely 1.29 ton/ha and relatively high value of the CV%, namely 4.54% were found.

Table 3.5 Descriptive statistics of seasonal weather elements and maize productivity (ton/ha) in the fourth agro-climatic zones of Egypt from 2009 to 2021

Statistic	SRAD	T_{max}	T_{min}	RH	Maize productivity
Minimum value	27.44	37.05	21.03	24.48	6.79
Maximum value	28.47	41.50	23.91	30.61	8.08
Range	1.04	4.45	2.88	6.12	1.29
Mean	27.98	38.84	22.27	27.49	7.49
Standard deviation	0.25	1.54	0.81	1.93	0.34
CV (%)	0.91	3.97	3.64	7.01	4.54

SRAD = solar radiation (MJ/m^2/day), T_{max} = maximum temperature (°C), T_{min} = minimum temperature (°C), RH = relative humidity (%)

The Fifth Agro-climatic Zone

The results in Table 3.6 revealed that the lowest values of the range between minimum and maximum values was found for seasonal minimum temperature, namely 2.86 °C and the lowest value of CV% was found for seasonal maximum temperature, namely 2.45 °C. Whereas, the highest values of the range between minimum and maximum values and CV% were found for seasonal relative humidity, namely 3.14% and 4.69%. These values were the lowest compared to those found in the rest of the agro-climatic zones. With respect to maize productivity, the highest value of the range between minimum and maximum values was 2.68 ton/ha and the highest value of the CV% was 14.94%.

The above results suggested that maize is most suitable to grow in the fourth agro-climatic zone, because there was less fluctuation in its productivity in this zone, compared to its value in the rest of the agro-climatic zones.

Table 3.6 Descriptive statistics of seasonal weather elements and maize productivity (ton/ha) in the fifth agro-climatic zones of Egypt from 2009 to 2021

Statistic	SRAD	T_{max}	T_{min}	RH	Maize productivity
Minimum value	25.57	39.35	22.77	17.09	4.14
Maximum value	28.56	42.62	25.63	20.23	6.83
Range	2.99	3.27	2.86	3.14	2.68
Mean	27.98	40.86	23.97	18.50	5.27
Standard deviation	0.75	1.00	0.81	0.87	0.79
CV (%)	2.67	2.45	3.38	4.69	14.94

SRAD = solar radiation (MJ/m^2/day), T_{max} = maximum temperature (°C), T_{min} = minimum temperature (°C), RH = relative humidity (%)

The Strength of the Relationship Between Maize Productivity and the Seasonal Weather Elements

The Strength of the Relationship Between Zone-Specific Maize Productivity and the Seasonal Weather Elements in the Agro-climatic Zones of Egypt

The First Agro-climatic Zone

The multiple linear regression analysis showed a significant relationship ($P < 0.05$) between seasonal weather elements and maize productivity in the first agro-climatic zone, where a R^2 value was 0.6877*, which means that 68.77% of the variation in maize productivity can be explained by the studied weather elements. The goodness of fit analysis showed that RMSE value between measured and predicted maize productivity was 0.2238 ton/ha and *d-stat* value was 0.8999.

Because high variability in the seasonal solar radiation existed in this zone, represented by higher CV% value (Table 3.2), seasonal solar radiation has a significant positive effect on maize productivity ($P < 0.05$) (Fig. 3.2a). The results also showed insignificant negative relationships between maximum and minimum temperatures and maize productivity with R^2 values 0.5612 and 0.6024, respectively (Fig. 3.2b,c).

The Second Agro-climatic Zone

The multiple linear regression analysis between maize productivity and the studied seasonal weather elements in the second agro-climatic zone showed a significant relationship ($P < 0.05$), where R^2 value was 0.7802*. Furthermore, the goodness of fit analysis indicated that RMSE value between measured and predicted maize productivity was 0.1969 ton/ha and *d-stat* value was 0.9345.

The second agro-climatic zone has high variability in the seasonal solar radiation and seasonal maximum temperature (Table 3.3), and both have a significant relationship with maize productivity in this zone. Furthermore, the seasonal solar radiation has a significant positive relationship ($P < 0.05$) with maize productivity (Fig. 3.3a), whereas a negative significant relationship between seasonal maximum temperature ($P < 0.05$) and maize productivity was found (Fig. 3.3b).

The Third Agro-climatic Zone

A significant relationship resulted from the multiple linear regression analysis between seasonal weather elements and maize productivity ($P < 0.05$) in the third agro-climatic zone was found, where R^2 value was 0.7124*. Furthermore, the goodness of fit analysis revealed a low RMSE value between measured and predicted

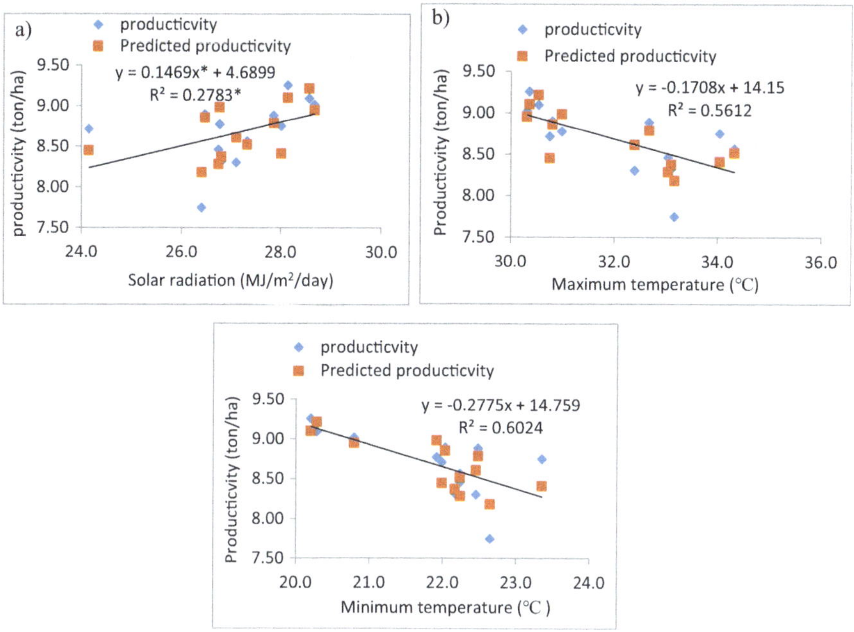

Fig. 3.2 Line fit between maize productivity and seasonal solar radiation (**a**), seasonal maximum temperature (**b**) and seasonal minimum temperatures (**c**) in the first agro-climatic zone

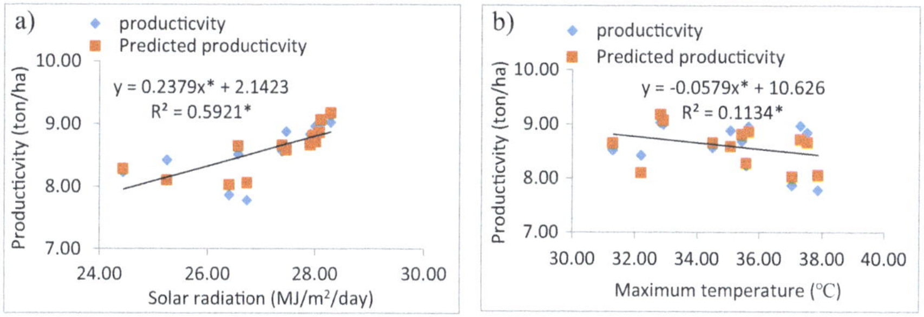

Fig. 3.3 Line fit between maize productivity and seasonal solar radiation, and maximum temperature in the second agro-climatic zone

maize productivity, namely 0.2871 ton/ha and high value of *d-stat* value, namely 0.9090.

The third agro-climatic zone has the lowest variability in the seasonal minimum temperature (Table 3.4) and the second lowest variability in the seasonal relative humidity (Table 3.4). Therefore, Fig. 3.4a showed a positive significant relationship ($P < 0.05$) between maize productivity and seasonal minimum temperature. Whereas, Fig. 3.4b showed a negative significant relationship ($P < 0.05$) between maize productivity and seasonal relative humidity.

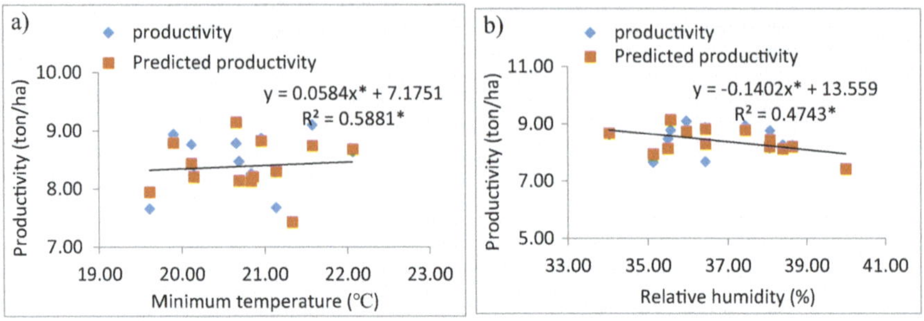

Fig. 3.4 Line fit between maize productivity and seasonal minimum temperature (**a**), and seasonal relative humidity (**b**) in the third agro-climatic zone

Fig. 3.5 Line fit between maize productivity and seasonal maximum temperature in the fourth agro-climatic zone

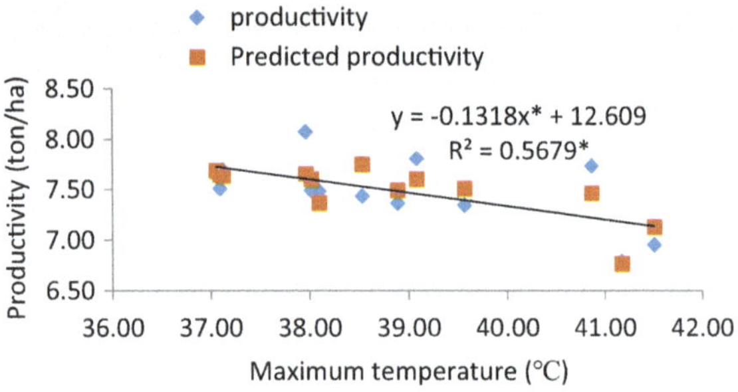

The Fourth Agro-climatic Zone

The obtained results from the multiple linear regression analysis indicated a significant relationship between the seasonal weather elements and maize productivity ($P < 0.05$) in the fourth agro-climatic zone, where R^2 value was 0.6290*. Additionally, the goodness of fit analysis showed that the values of RMSE and *d-stat* between measured and predicted maize productivity were 0.2071 ton/ha and 0.8742, respectively.

A significant negative relationship between seasonal maximum temperature ($P < 0.05$) and maize productivity was found in the fourth agro-climatic zone (Fig. 3.5).

The Fifth Agro-climatic Zone

The multiple linear regression analysis showed a significant relationship between the seasonal weather elements and maize productivity ($P < 0.05$) in the fifth agro-climatic zone, where R^2 value was 0.5963. The results of the goodness of fit analysis showed that RMSE value between measured and predicted maize productivity was 0.4997 ton/ha and *d-stat* value was 0.8577.

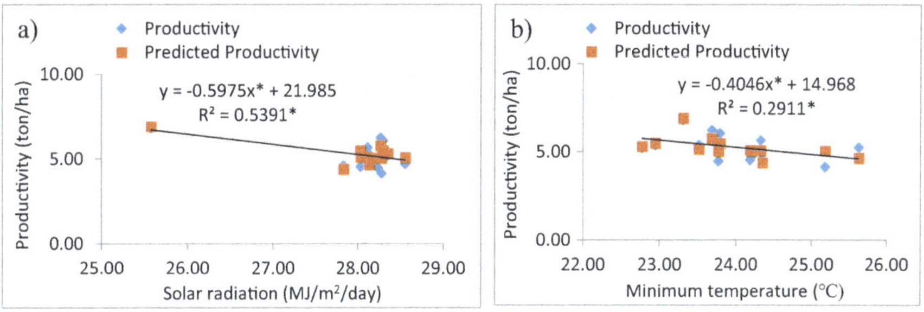

Fig. 3.6 Line fit between maize productivity and seasonal solar radiation (**a**) and seasonal minimum temperature (**b**) in the fifth agro-climatic zone

Figure 3.6a showed a significant negative relationship between the seasonal solar radiation and maize productivity ($P < 0.05$). Morsy et al. (2017) stated that, in Egypt, solar radiation decreases gradually from south to north according to the apparent position of the sun and reaches its maximum value in the summer season, which could explain the relatively high contribution of solar radiation in maize productivity in the fifth agro-climatic zone. Furthermore, a significant negative relationship between the seasonal minimum temperature and maize productivity ($P < 0.05$) was found (Fig. 3.6b).

The Strength of the Relationship Between Maize Productivity and the Seasonal Weather Elements in the Country of Egypt

A highly significant relationship between the studied seasonal weather elements and maize productivity ($P < 0.01$) was found for the country of Egypt, where R^2 value was 0.8163**. The goodness of fit analysis showed that RMSE value between measured and predicted maize productivity was 0.1385 ton/ha and *d-stat* value was 0.9997.

Furthermore, Fig. 3.7a showed a significant negative relationship between the seasonal maximum temperature and maize productivity ($P < 0.05$), as well as a highly significant negative relationship between maize productivity and seasonal relative humidity ($P < 0.01$) (Fig. 3.7b).

The above results suggest that the main seasonal weather elements affecting maize productivity in the five agro-climatic zones of Egypt are seasonal solar radiation in the first, second and fifth agro-climatic zones. The seasonal maximum temperature is also an affecting factor on maize productivity in first, second and fourth agro-climatic zones. The seasonal minimum temperature has an effect on maize productivity in the first, third and fifth agro-climatic zones. Furthermore, seasonal relative humidity also has an effect in the third agro-climatic zone only. Nevertheless, the results over

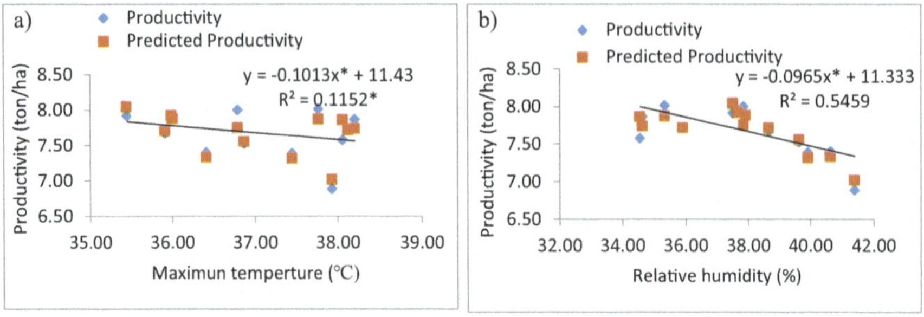

Fig. 3.7 Line fit between maize productivity and seasonal maximum temperature (**a**), and seasonal relative humidity (**b**) for the country of Egypt

Egypt showed that the most affecting weather elements on maize productivity were seasonal maximum temperature and seasonal relative humidity.

The Future Climate of Egypt

The future values of yearly maximum and minimum temperatures and relative humidity of the two future climate scenarios, namely SSP1-2.6 and SSP5-8.5 were graphed to show the expected pattern of these climate elements. As it was stated earlier, the SSP1-2.6 assumes that increasing sustainability will severely cut the global emissions. As a result, it is expected to cause a slow rise in the seasonal maximum temperature in Egypt, and in its agro-climatic zones. Whereas, the SSP5-8.5 assumed that the misuse of fossil fuel resources will be intensified and that will cause fast rise in the seasonal maximum temperature in Egypt.

The Future Climate of the Agro-climatic Zones of Egypt

The First Agro-climatic Zone

In the first agro-climatic zone, Fig. 3.8 showed that both future scenarios SSP1-2.6 and SSP5-8.5 will cause a rise in the three studied climate elements, in the time interval between 2024 and 2094. Slight differences in the values of maximum temperature between both scenarios (Fig. 3.8a) were found between 2024 and 2044. SSP1-2.6 will show a slight peak occur in 2054 and then was almost leveled until 2094. Whereas, SSP5-8.5 will show an increasing trend until 2094. In 2094, SSP5-8.5 projected a higher value of maximum temperature than the value under SSP1-2.6 by 3.1 °C. Line fit for maximum temperature projected by SSP5-8.5 showed a higher value of R^2, namely 0.9902, whereas the value of R^2 was 0.4686 under SSP1-2.6 (Fig. 3.8a).

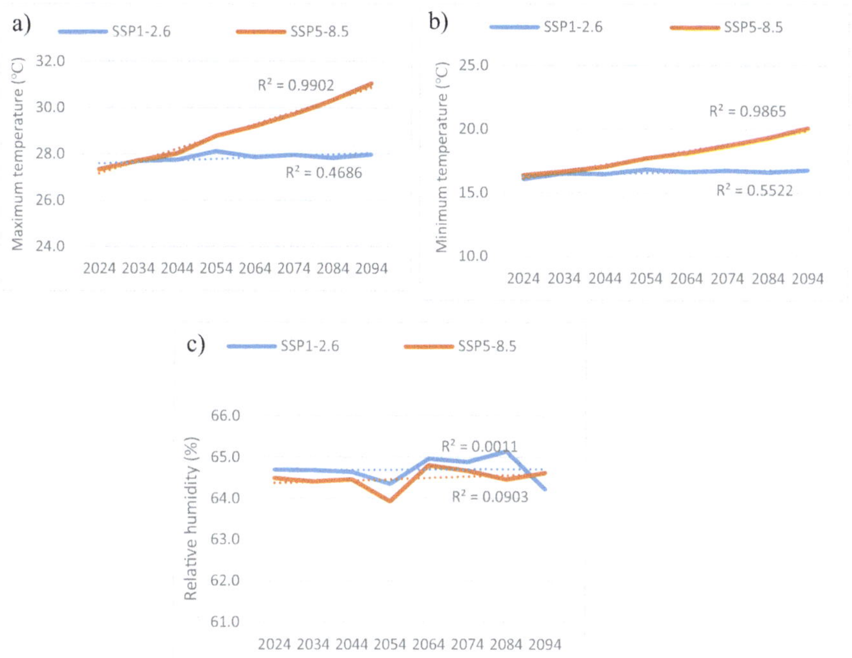

Fig. 3.8 The projected trend of maximum temperature (**a**), minimum temperature (**b**) and relative humidity (**c**) developed by SSP1-2.6 and SSP5-8.5 future climate scenarios in the first agro-climatic zone of Egypt. *Source of data* https://climateknowledgeportal.worldbank.org/

Similar trend will be found for minimum temperature (Fig. 3.8b), where SSP5-8.5 projected a higher value than SSP1-2.6 by 3.3 °C in 2094. Furthermore, line fit for minimum temperature projected by SSP5-8.5 showed a higher value of R^2, namely 0.9865, whereas the value of R^2 was 0.5522 under SSP1-2.6.

The projected values of relative humidity by the two future scenarios were almost parallel from 2024 to 2044. Both scenarios projected a decline in the relative humidity, however the decline was deeper under SSP5-8.5, compared to SSP1-2.6. A rise and slight decline, followed by another rise and sharp declined in SSP1-2.6 between 2064 and 2094. A rise in relative humidity by 2064 projected by SSP5-8.5 then decline in 2084 and slight rise will occur. Furthermore, SSP5-8.5 projected a higher value of relative humidity than SSP1-2.6 by 0.4% in 2094. Low value of R^2 resulted from line fit for relative humidity projected by SSP5-8.5 was found, namely 0.0903, whereas the value of R^2 was 0.0011 under SSP1-2.6 (Fig. 3.8c).

The Second Agro-climatic Zone

Similar trends of maximum temperature to the ones exist in the first agro-climatic zone will be found in the second agro-climatic zone (Fig. 3.9a) and minimum temperature (Fig. 3.9b). However, in 2094, SSP5-8.5 projected a higher value than SSP1-2.6 by 3.5 °C in maximum temperature. Line fit for maximum temperature projected by SSP5-8.5 showed higher value of R^2, namely 0.9906 (Fig. 3.9a), whereas the value R^2 was 0.4608 under SSP1-2.6 (Fig. 3.9a). Similar trend will be found for minimum temperature, but with higher difference between the two future scenarios. Furthermore, SSP5-8.5 projected a higher value than SSP1-2.6 by 3.8 °C in 2094. Furthermore, the value of R^2 resulted from the line fit of minimum temperature was 0.9843 under SSP5-8.5 and it was 0.5556 under SSP1-2.6.

With respect of relative humidity, a completely different tends than the ones projected in the first agro-climatic zone will be found. The projected values of relative humidity by SSP1-2.6 will slightly decline in 2034 and slightly rise in 2044 and will sharply decline in 2054. It will sharply rise in 2064 and continue to rise until 2084 then it will decline in 2094. Different trend will be found for SSP5-8.5, where the relative humidity will be lower than the projected in SSP1-2.6 and it will sharply decline until 2054 then sharply rise in 2064. It will sharply decline in 2084 then

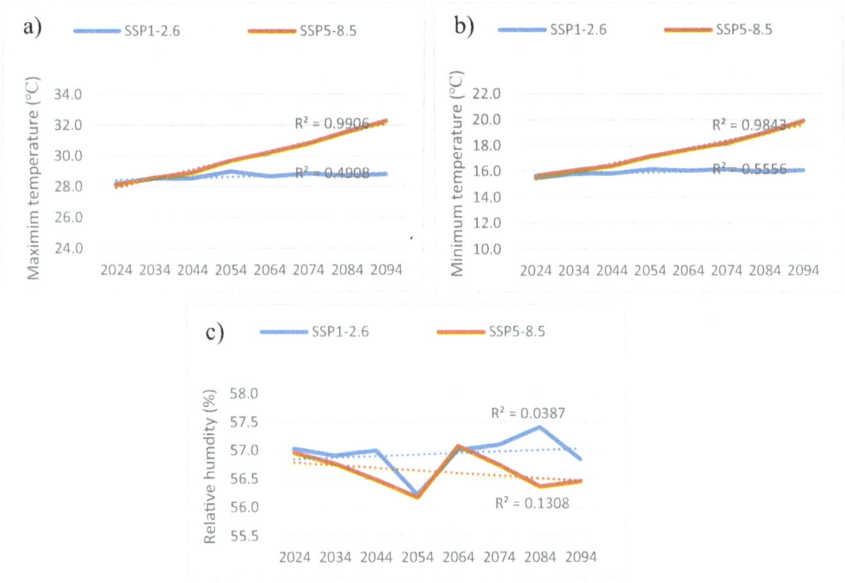

Fig. 3.9 The projected trend of maximum temperature (**a**), minimum temperature (**b**) and relative humidity (**c**) developed by SSP1-2.6 and SSP5-8.5 future climate scenarios in the second agro-climatic zone of Egypt. *Source of data* https://climateknowledgeportal.worldbank.org/

slightly rise until 2094. Furthermore, SSP5-8.5 projected a lower value than SSP1-2.6 by 0.4% in 2094 (Fig. 3.9c). Low value of R^2 resulted from line fit of SSP5-8.5 was 0.1308 and it was 0.0387 for SSP1-2.6.

The Third Agro-climatic Zone

In the third agro-climatic zone, similar trends to the ones exist in the first and second agro-climatic zones will be found for maximum temperature (Fig. 3.10a) and minimum temperature (Fig. 3.10b). However, in 2094, SSP5-8.5 projected a higher value of maximum temperature than the projected by SSP1-2.6 by 3.8 °C. Similar trend will be found for minimum temperature, but with higher difference between the two future scenarios. Furthermore, SSP5-8.5 projected a higher value of minimum temperature than the values obtained by SSP1-2.6 by 4.4 °C in 2094.

The relative humidity will follow similar tends like the ones projected in the second agro-climatic zone. However, SSP5-8.5 projected a lower value than the obtained by SSP1-2.6 in 2094 by 1.1% (Fig. 3.10c).

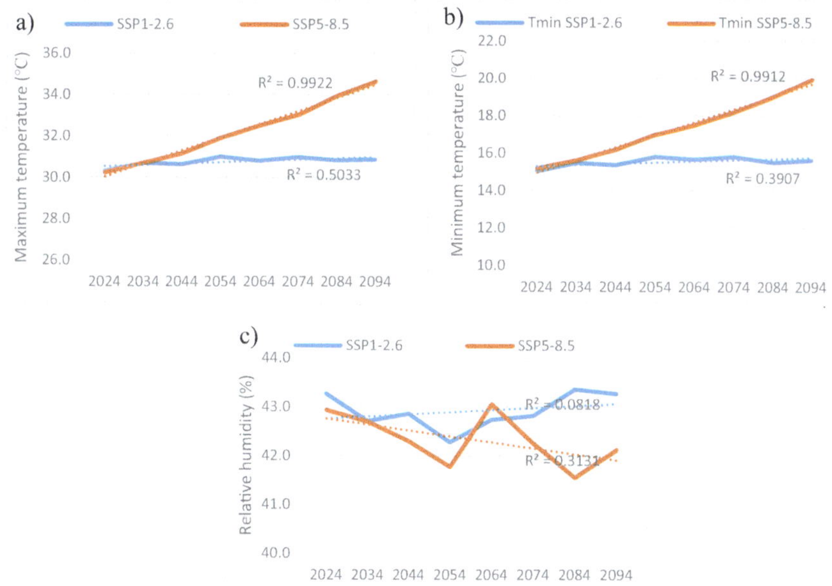

Fig. 3.10 The projected trend of maximum temperature (**a**), minimum temperature (**b**) and relative humidity (**c**) developed by SSP1-2.6 and SSP5-8.5 future climate scenarios in the third agro-climatic zone of Egypt. *Source of data* https://climateknowledgeportal.worldbank.org/

The Fourth Agro-climatic Zone

Maximum temperature (Fig. 3.11a) and minimum temperature (Fig. 3.11b) in the fourth agro-climatic zone will follow similar trends to the ones found in the first, second and third agro-climatic zones. However, in 2094, SSP5-8.5 projected a higher value than SSP1-2.6 by 3.9 °C. Similar trend will be found for minimum temperature, but with higher difference between the two future scenarios, where SSP5-8.5 projected a higher value than SSP1-2.6 by 4.5 °C in 2094.

The relative humidity will follow similar trend like those projected in the third agro-climatic zone. The relative humidity values projected by SSP1-2.6 will highly decline in 2034 and then will rise in 2064. It will decline again in 2054 and became steady until 2074. It will rise then decline in 2084 and 2094, respectively. Different trend will be found for SSP5-8.5, where the relative humidity will sightly decline from 2024 to 2034 than sharply decline until 2054. It will sharply rise and decline between 2064 and 2084 respectively then it will slightly rise in 2094. However, SSP5-8.5 projected a lower value than the obtained by SSP1-2.6 in 2094 by 1.2% (Fig. 3.11c).

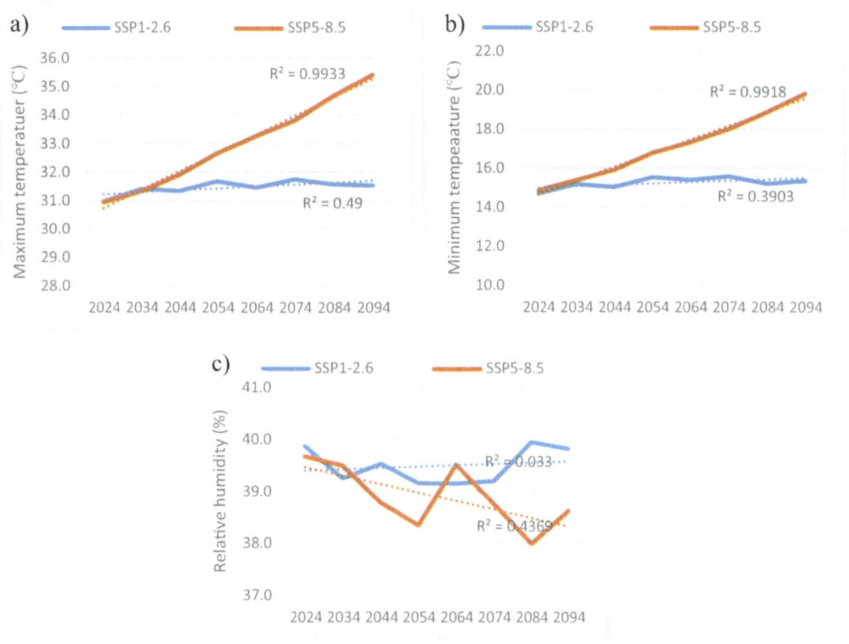

Fig. 3.11 The projected trend of maximum temperature (**a**), minimum temperature (**b**) and relative humidity (**c**) developed by SSP1-2.6 and SSP5-8.5 future climate scenarios in the fourth agro-climatic zone of Egypt. *Source of data* https://climateknowledgeportal.worldbank.org/

The Fifth Agro-climatic Zone

In the fifth agro-climatic zone, maximum temperature (Fig. 3.12a) and minimum temperature (Fig. 3.12b) will follow similar trends to those found in the first, second, third and fourth agro-climatic zones. However, in 2094, SSP5-8.5 projected a higher value of maximum temperature than SSP1-2.6 by 4.0 °C. Similar trend will be found for minimum temperature, but with higher difference between the two future scenarios, where SSP5-8.5 projected a higher value than SSP1-2.6 by 4.8 °C in 2094.

The relative humidity values projected by SSP5-8.5 in the fifth agro-climatic zone was higher in 2024 than its counterpart value projected by SSP1-2.6. The relative humidity values projected by SSP1-2.6 will decline in 2034 and then will rise in 2054. It will decline again in 2064, rise again in 2074 and became steady until 2094. Different trend will be found for SSP5-8.5, where the relative humidity will gradually decline from 2024 to 2074 than sharply decline until 2084 and 2094. However, SSP5-8.5 projected a lower value than SSP1-2.6 in 2094 by 1.1% (Fig. 3.12c).

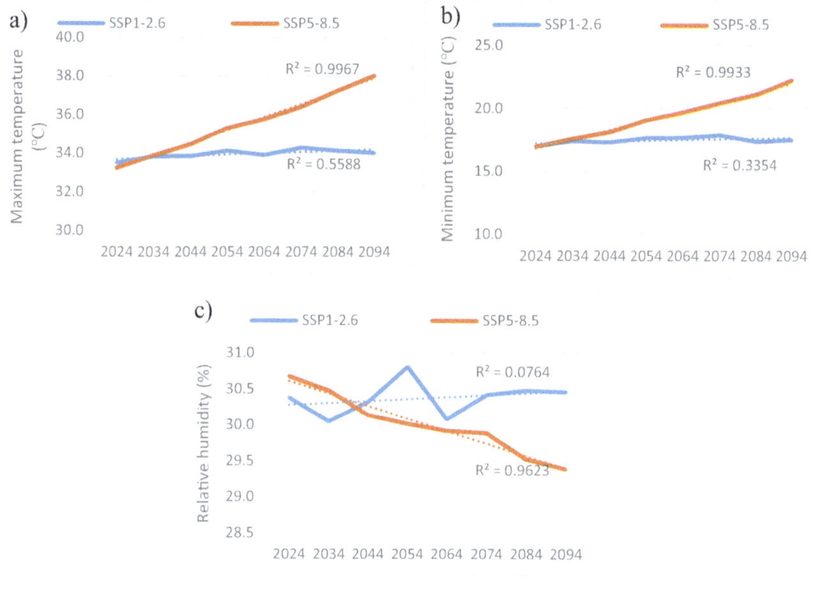

Fig. 3.12 The projected trend of maximum temperature (**a**), minimum temperature (**b**) and relative humidity (**c**) developed by SSP1-2.6 and SSP5-8.5 future climate scenarios in the fifth agro-climatic zone of Egypt. *Source of data* https://climateknowledgeportal.worldbank.org/

The Projected Climate of the Country of Egypt

Similar trends to those exist in the first, second, third, fourth and fifth agro-climatic zones for maximum temperature (Fig. 3.13a) and minimum temperature (Fig. 3.13b) will be found in the country of Egypt. However, in 2094, SSP5-8.5 projected a higher value of maximum temperature than its value under SSP1-2.6 by 3.8 °C. Similar trend will be found for minimum temperature, but with higher difference between the two future scenarios, where SSP5-8.5 projected a higher value than SSP1-2.6 by 4.4 °C in 2094.

Figure 3.13c showed that the relative humidity values projected by both SSP5-8.5 and SSP1-2.6 will be very close between 2024 and 2034. But, the projected values by SSP1-2.6 will rise in 2044 and then will decline in 2054. A gradual rise will continue until 2094. Different trend will be found for SSP5-8.5, where the relative humidity continues to decline until 2054 then rise 2064, then sharply decline until 2084 and rise again in 2094. Furthermore, SSP5-8.5 projected a lower value than SSP1-2.6 in 2094 by 1.1%.

The results in Table 3.7 summarized the expected climate of Egypt under both future scenarios. The two scenarios projected a gradual increase in the maximum temperature from the first agro-climatic zone to the fifth agro-climatic zone under both future scenarios. Similar trend can be noticed for minimum temperature, where it was gradually increase toward the fifth agro-climatic zone. The opposite trend can

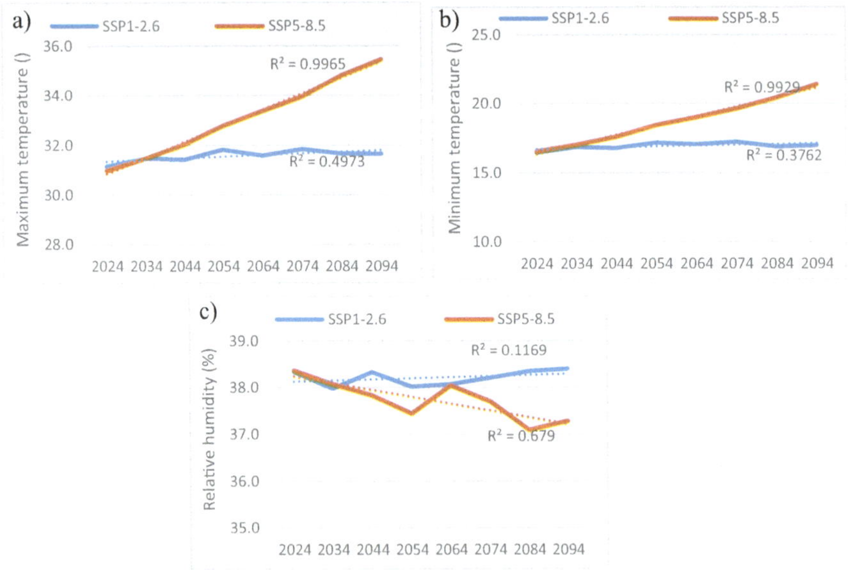

Fig. 3.13 The projected trend of maximum temperature (**a**), minimum temperature (**b**) and relative humidity (**c**) developed by SSP1-2.6 and SSP5-8.5 future climate scenarios for the country of Egypt. *Source of data* https://climateknowledgeportal.worldbank.org/

Table 3.7 The projected values of yearly maximum and minimum temperatures and relative humidity averaged from 2024 to 2094 in the agro-climatic zones of Egypt

Agro-climatic zones	T_{max} (°C)		T_{min} (°C)		RH (%)	
	SSP1-2.6	SSP5-8.5	SSP1-2.6	SSP5-8.5	SSP1-2.6	SSP5-8.5
First agro-climatic zone	27.82	29.03	16.60	18.04	64.71	64.49
Second agro-climatic zone	28.67	30.04	15.96	17.53	56.94	56.63
Third agro-climatic zone	30.77	32.28	15.54	17.36	42.91	42.34
Fourth agro-climatic zone	31.47	33.02	15.27	17.15	39.50	38.91
Fifth agro-climatic zone	33.96	35.56	17.50	19.45	30.38	30.01
Egypt	31.58	33.12	16.95	18.80	38.21	37.74

T_{max} = yearly maximum temperature (°C), T_{min} = yearly minimum temperature (°C), RH = yearly relative humidity (%)
Source of data https://climateknowledgeportal.worldbank.org/

be observed for relative humidity, where it will gradually decrease toward the fifth agro-climatic zone.

Projection of Maize Productivity in the Future

The seasonal weather elements calculated for maize growing season, namely seasonal maximum and minimum temperature and seasonal relative humidity were regressed on maize productivity in the time period from 2009 to 2021 to develop a zonal-specific prediction equation to be used in the projection of maize productivity in the future using the two climate scenarios, namely SSP1-2.6 and SSP5-8.5 obtained from CMIP6 multi-model ensemble. The developed prediction equations are included in Table 3.8. The table also revealed that the developed prediction equations in the first agro-climatic zone was found significant ($P < 0.10$), with low R^2 value of 0.5120, high *d-stat* value of 0.8131 and low value of RMSE, namely 0.2814 ton/ha. The developed prediction equations in the second, the third and the fourth agro-climatic zones were found significant ($P < 0.05$). R^2 values were 0.6316, 0.6386 and 0.6191, respectively, the *d-stat* values were 0.8806, 0.8808 and 0.8697, respectively and the values of RMSE were 0.2550, 0.3218 and 0.2099 ton/ha, respectively.

Although the developed prediction equation in the fifth agro-climatic zone was found insignificant, we used it in the prediction just to show the trend of the future maize productivity under the future climate scenarios. The value of R^2 value was 0.2225, *d-stat* value was 0.5492 and the value of RMSE, namely 0.6936 ton/ha. The developed prediction equation for Egypt was found highly significant ($P < 0.01$), with a high value of R^2, namely 0.8163, high value of *d-stat,* namely 0.9996 and a low value of RMSE, namely 0.2814 ton/ha (Table 3.8).

Table 3.8 The prediction equations for maize productivity in the agro-climatic zones of Egypt using weather data from 2009 to 2021

Zone	Prediction equation	R^2	*d-stat*	RMSE[1]	Sig. F
Zone 1	$y\hat{} = 14.4661 - 0.0662 * T_{max} - 0.4975 * T_{min} + 0.0707 * RH$	0.5120	0.8131	0.2814	0.0832
Zone 2	$y\hat{} = 13.2025 - 0.0953 * T_{max} - 0.1248 * T_{min} - 0.0272 * RH$	0.6316	0.8806	0.2550	0.0241
Zone 3	$y\hat{} = -6.3539 + 0.8822 * T_{max} - 1.2912 * T_{min} + 0.1901 * RH$	0.6386	0.8809	0.3218	0.0222
Zone 4	$y\hat{} = 12.3548 - 0.2702 * T_{max} + 0.2069 * T_{min} - 0.0704 * RH$	0.6191	0.8697	0.2099	0.0279
Zone 5	$y\hat{} = 15.2213 + 0.1639 * T_{max} - 0.6201 * T_{min} - 0.1813 * RH$	0.2225	0.5492	0.6936	0.4969
Egypt	$y\hat{} = 17.1299 - 0.1523 * T_{max} - 0.1035 * T_{min} - 0.1199 * RH$	0.8163	0.9996	0.1612	0.0012

T_{max} = seasonal maximum temperature (°C), T_{min} = seasonal minimum temperature (°C), RH = seasonal relative humidity (%), [a]ton/ha

Maize Productivity Projected by SSP1-2.6 Future Scenario in Egypt

The projected values of minimum and maximum temperature and relative humidity were applied in the prediction equations of the agro-climatic zone of Egypt (presented in Table 3.8) to project maize productivity in the future. The projected maize productivity was compared to a baseline value, which is the zonal average productivity value in 2014 in each agro-climatic zone.

Maize Productivity Projected in the Agro-climatic Zones of Egypt

The First Agro-climatic Zone

Table 3.9 showed the future seasonal values of maximum and minimum temperatures and relative humidity in the first agro-climatic zone. It can be notice from the table that the values of maximum and maximum temperatures will be gradually increase from 2020 to 2039 until 2080–2099 with the highest value will occur in 2094. The mean values of the maximum temperature will reach 34.71 °C with a range between minimum and maximum values will reach 2.94 °C. The mean values of the seasonal minimum temperature will reach 21.19 °C with a range between minimum and maximum values will reach 0.51 °C. Whereas, the seasonal values of relative humidity will be gradually increase until 2040–2059, then it will slightly decrease until 2080–2099. The mean value of seasonal relative humidity will be 59.63% with a range between minimum and maximum values will be 11.13%.

The results in Table 3.9 also showed that the projected values of maize productivity will be lower than the baseline value, with the lowest value will exist in 2020–2039, namely 5.52 ton/ha. The mean value of projected maize productivity will be 5.84 ton/ha with a range between minimum and maximum values of 0.56 ton/ha.

Because SSP1-2.6 is considered as optimistic future scenario, where it assumes that global emissions will be declined and reach net-zero after 2050, the reduction in the average maize productivity was lower in the period from 2040 to 2059, than its value in 2020–2039. However, the reduction in average maize productivity will stabilize to be 33% until 2099. The lowest percentage of reduction between baseline and projected productivity will occur in 2040–2059 and the highest percentage of reduction will occur in 2020–2039, with an average of 33%, and a range between minimum and maximum values of 4%.

Table 3.9 Projected seasonal climate elements and maize productivity in the future under SSP1-2.6 future climate scenario in the first agro-climatic zone

	T_{max} (°C)	T_{min} (°C)	RH (%)	Projected maize productivity (ton/ha)	Percentage of reduction (%)
2020–2039	33.07	20.87	51.31	5.52	37
2040–2059	33.86	21.23	62.44	6.08	31
2060–2079	35.91	21.38	62.40	5.87	33
2080–2099	36.01	21.28	62.36	5.91	33
Mean	34.71	21.19	59.63	5.84	33
Range	2.94	0.51	11.13	0.56	4

T_{max} = maximum temperature (°C), T_{min} = minimum temperature (°C), RH = relative humidity (%)

Baseline maize productivity = 8.90 ton/ha

In a previous study by Morsy (2015), he projected a decline in maize productivity in 2030 by 9%, compared to its value in 2010 in the first agro-climatic zone.

The Second Agro-climatic Zone

The results in Table 3.10 showed that the values of seasonal maximum temperature will gradually increase until 2040–2059 and then it will decline until 2080–2099. The mean values of the seasonal maximum temperature will reach 34.83 °C with a range between minimum and maximum values will reach 1.65 °C. Whereas, the values of seasonal minimum temperature will gradually increase in 2020–2039 until 2060–2079 and then slightly decline in 2080–2099. The mean values of seasonal minimum temperature will reach 21.42 °C with a range between minimum and maximum values will reach 0.57 °C. Furthermore, the seasonal values of relative humidity will be gradually decline until 2060–2079, then it will rise in 2080–2099. The mean value of seasonal relative humidity will be 53.83% with a range between minimum and maximum values will be 11.53%.

The projected values of maize productivity will be lower than the baseline value, with a mean value of 5.75 ton/ha and a range between minimum and maximum values of 0.37 ton/ha. The same trend observed in the first agro-climatic zone was found in the second climatic zone, where the reduction in the average maize productivity will be lower in the period from 2040 to 2059, than its value in 2020–2039, namely 38% versus 36%, respectively. The reduction in the average maize productivity will be then stabilize to be 34% until 2094. The average value of the percentage of reduction between baseline and projected maize productivity will be 36%, with a range between minimum and maximum values of 4% (Table 3.10).

Morsy (2015) projected a decline in maize productivity in the second agro-climatic zone in 2030 by 12%, compared to its value in 2010. Whereas, Ouda et al. (2012)

Table 3.10 Projected climate elements and maize productivity in the future under SSP1-2.6 future climate scenario in the second agro-climatic zone

	T_{max} (°C)	T_{min} (°C)	RH (%)	Projected maize productivity (ton/ha)	Percentage of reduction (%)
2020–2039	35.38	21.04	62.39	5.51	38
2040–2059	35.75	21.49	51.03	5.73	36
2060–2079	34.11	21.61	50.87	5.88	34
2080–2099	34.10	21.55	51.05	5.88	34
Mean	34.83	21.42	53.83	5.75	36
Range	1.65	0.57	11.53	0.37	4

T_{max} = maximum temperature (°C), T_{min} = minimum temperature (°C), RH = relative humidity (%)

Baseline maize productivity = 8.96 ton/ha

projected a reduction in maize productivity by 26% and Taha (2012) projected a decline in maize productivity by 38% in the same agro-climatic zone in 2030.

The Third Agro-climatic Zone

In the third agro-climatic zone, the values of seasonal maximum and minimum temperatures will gradually increase until 2060–2079 and then it will decline until 2080–2099. The mean values of the seasonal maximum and minimum temperatures will reach 38.54 and 22.40 °C, respectively with a range between minimum and maximum values will reach 1.83 and 0.59 °C, respectively. Whereas, the seasonal values of relative humidity will be gradually decline until 2080–2099. The mean value of seasonal relative humidity will be 32.71% with a range between minimum and maximum values will be 0.54% (Table 3.11).

Higher losses in the productivity of maize in the third agro-climatic zone will occur. The lowest value of future maize productivity will exist in 2020–2039, namely 4.38 ton/ha, with a percentage of reduction of 50%. Whereas, the highest value of future maize productivity will exist in 2080–2099, namely 5.25 ton/ha, with a percentage of reduction of 40%. The mean value of future maize will be 4.95 ton/ha with a range between minimum and maximum values of 0.87 ton/ha. The lowest percentage of reduction between baseline and projected productivity will occur in 2080–2099 and the highest percentage will occur in 2020–2039, with an average of 44%, and a range between minimum and maximum values of 10% (Table 3.11).

Ouda et al. (2009) projected reduction in maize productivity in 2030 by 47% and Moneir (2012) projected a decline in maize productivity by 46% in the third agro-climatic zone in 2030.

Table 3.11 Projected climate elements and maize productivity in the future under SSP1-2.6 future climate scenario in the third agro-climatic zone

	T_{max} (°C)	T_{min} (°C)	RH (%)	Projected maize productivity (ton/ha)	Percentage of reduction (%)
2020–2039	37.25	22.01	33.05	4.38	50
2040–2059	38.88	22.54	32.54	5.04	43
2060–2079	39.08	22.60	32.51	5.13	41
2080–2099	38.97	22.46	32.74	5.25	40
Mean	38.54	22.40	32.71	4.95	44
Range	1.83	0.59	0.54	0.87	10

T_{max} = maximum temperature (°C), T_{min} = minimum temperature (°C), RH = relative humidity (%)

Baseline maize productivity = 8.77 ton/ha

Table 3.12 Projected climate elements and maize productivity in the future under SSP1-2.6 future climate scenario in the fourth agro-climatic zone

	T_{max} (°C)	T_{min} (°C)	RH (%)	Projected maize productivity (ton/ha)	Percentage of reduction (%)
2020–2039	39.31	22.20	29.06	4.48	42
2040–2059	39.76	22.69	28.67	4.49	42
2060–2079	39.85	22.81	28.74	4.49	42
2080–2099	39.70	22.58	28.80	4.47	42
Mean	39.66	22.57	28.82	4.48	42
Range	0.54	0.62	0.39	0.01	0

T_{max} = maximum temperature (°C), T_{min} = minimum temperature (°C), RH = relative humidity (%)
Baseline maize productivity = 7.70 ton/ha

The Fourth Agro-climatic Zone

In the fourth agro-climatic, the values of seasonal maximum and minimum temperatures will slightly increase until 2060–2079 and then it will decline until 2080–2099. Consequently, the range between minimum and maximum values of temperatures will reach 0.54 and 0.62 °C, respectively, with the mean values will reach 39.66 and 22.57 °C, respectively. Whereas, the seasonal values of relative humidity will be slightly decline until 2040–2059 then it will rise until 2080–2099. The mean value of seasonal relative humidity will be 28.82% with a range between minimum and maximum values will be 0.39%. The observed low values of the range between minimum and maximum values of the studied climatic elements (Table 3.12) implies that there will be a relative stability in the climate between 2020 and 2099 in the fourth agro-climatic zone.

Low variability in the projected values of maize productivity, which will result in low value of the range between minimum and maximum values, namely 0.01 ton/ha with a mean value of 4.48 ton/ha was observed. Accordingly, the percentage of reduction between baseline and projected productivity will be 42%, in all the studied time intervals (Table 3.12).

The Fifth Agro-climatic Zone

In the fifth agro-climatic zone (Table 3.13), the values of seasonal maximum and minimum temperatures will highly rise in 2040–2059, compared to 2020–2039. Then, it will slightly rise until 2060–2079 and it will slightly decline until 2080–2099. The mean values of the seasonal maximum and minimum temperatures will reach 41.86 and 25.03 °C, respectively with a range between minimum and maximum values will reach 2.21 and 0.51 °C, respectively. Whereas, the seasonal values of relative humidity will be gradually decline until 2060–2079 and then it will rise until

Table 3.13 Projected climate elements and maize productivity in the future under SSP1-2.6 future climate scenario in the fifth agro-climatic zone

	T_{max} (°C)	T_{min} (°C)	RH (%)	Projected maize productivity (ton/ha)	Percentage of reduction (%)
2020–2039	40.26	24.68	21.37	2.45	54
2040–2059	42.42	25.19	21.14	2.52	53
2060–2079	42.48	25.32	21.05	2.46	54
2080–2099	42.29	24.95	21.14	2.65	51
Mean	41.86	25.03	21.17	2.52	53
Range	2.21	0.51	0.32	0.19	3

T_{max} = maximum temperature (°C), T_{min} = minimum temperature (°C), RH = relative humidity (%)
Baseline maize productivity = 5.37 ton/ha

2080–2099. The mean value of seasonal relative humidity will be 21.17% with a range between minimum and maximum values will be 0.32% (Table 3.13). Thus, low variability in the value of seasonal minimum temperature and seasonal relative humidity will be found.

Low variability in maize productivity will exist in the fifth agro-climatic zone, where the value of the range between minimum and maximum values was 0.19 ton/ha. Furthermore, the mean value of maize productivity will be 2.52 ton/ha. The lowest percentage of reduction between baseline and projected productivity will occur in 2080–2099, with an average of 53%, and a range between minimum and maximum values of 3% (Table 3.13).

The Country of Egypt

The expected behavior of the three studied climate elements will be similar to that of the fifth agro-climatic zone. The values of seasonal maximum and minimum temperatures will highly rise in 2040–2059, compared to 2020–2039. Then, it will slightly rise until 2060–2079 and it will slightly decline until 2080–2099. The mean values of the seasonal maximum and minimum temperatures will reach 38.41 and 23.84 °C, respectively with a range between minimum and maximum values will reach 4.82 and 0.09 °C, respectively. Whereas, the seasonal values of relative humidity will be gradually decline until 2080–2099 and then it will rise in 2080–2099. The mean value of seasonal relative humidity will be 29.09% with a range between minimum and maximum values will be 0.41% (Table 3.14).

The lowest value of future maize productivity will exist in 2060–2079. The projected values of maize productivity will be lower than the baseline value, with a mean value of 5.32 ton/ha and a range between minimum and maximum values of 0.74 ton/ha. The lowest percentage of reduction between baseline and projected

Table 3.14 Projected climate elements and maize productivity in the future under SSP1-2.6 future climate scenario averaged over Egypt

	T_{max} (°C)	T_{min} (°C)	RH (%)	Projected maize productivity (ton/ha)	Percentage of reduction (%)
2020–2039	34.86	23.48	29.35	5.87	26
2040–2059	39.57	23.96	29.04	5.14	35
2060–2079	39.68	24.06	28.95	5.13	35
2080–2099	39.55	23.85	29.02	5.16	35
Mean	38.41	23.84	29.09	5.32	33
Range	4.82	0.09	0.41	0.74	9

T_{max} = maximum temperature (°C), T_{min} = minimum temperature (°C), RH = relative humidity (%)
Baseline maize productivity = 7.94 ton/ha

productivity will occur in 2020–2039, with an average of 33%, and a range between minimum and maximum values of 9% (Table 3.14).

Maize Productivity Projected by SSP5-8.5 Future Scenario in Egypt

Maize Productivity Projected in the Agro-climatic Zones of Egypt

The First Agro-climatic Zone

Table 3.15 showed the future seasonal values of maximum and minimum temperature and relative humidity in the first agro-climatic zone under the assumption of SSP5-8.5 future climate scenario. The table revealed that there will be a progressive rise in the values of maximum and minimum temperatures from 2020 to 2039 until 2080–2099, where the highest value will occur in 2080–2099. The mean values of the seasonal maximum temperature will reach 35.31 °C with a range between minimum and maximum values will reach 3.94 °C. The mean values of the seasonal minimum temperature will reach 22.80 °C with a range between minimum and maximum values will reach 3.78 °C. Whereas, the seasonal values of relative humidity will be gradually decline until 2080–2099. The mean value of seasonal relative humidity will be 62.01% with a range between minimum and maximum values will be 1.03%. Thus, high variability in the value of seasonal maximum and minimum temperatures and relatively low variability in the value of seasonal relative humidity will occur under this scenario.

Table 3.15 Projected seasonal climate elements and maize productivity in the future under SSP5-8.5 future climate scenario in the first agro-climatic zone

	T_{max} (°C)	T_{min} (°C)	RH (%)	Projected maize productivity (ton/ha)	Percentage of reduction (%)
2020–2039	33.44	21.06	62.42	6.19	30
2040–2059	34.55	22.07	62.47	5.62	37
2060–2079	35.89	23.27	61.73	4.88	45
2080–2099	37.38	24.82	61.44	3.99	55
Mean	35.31	22.80	62.01	5.17	42
Range	3.94	3.76	1.03	2.20	25

T_{max} = maximum temperature (°C), T_{min} = minimum temperature (°C), RH = relative humidity (%)
Baseline maize productivity = 8.90 ton/ha

The results also revealed that the increasing trend in the values of seasonal maximum and minimum temperatures from 2020–2039 to 2080–2099 will result in a decreasing trend in maize productivity in the future. The projected values of maize productivity will be lower than the baseline value, with the lowest value will exist in 2020–2039, namely 6.19 ton/ha. The mean value of projected maize productivity will be 5.17 ton/ha with a range between minimum and maximum values of 2.20 ton/ha. The lowest percentage of reduction between baseline and projected productivity will occur in 2020–2039 and the highest percentage will occur in 2080–2099, with an average of 42%, and a range between minimum and maximum values of 25% (Table 3.15).

The Second Agro-climatic Zone

Similar trends to those found in the first agro-climatic zone will be observed in the second agro-climatic zone for the values of seasonal maximum and minimum temperatures, as well as seasonal relative humidity (Table 3.16). The values of seasonal maximum and minimum temperatures will gradually rise until 2080–2099. The values of the seasonal relative humidity will gradually decline until 2080–2099. The average value of seasonal maximum and minimum temperatures will reach 37.43 and 23.23 °C with a range between minimum and maximum values will reach 4.21 and 4.33 °C. Furthermore, the mean value of seasonal relative humidity will be 50.08% with a range between minimum and maximum values will be 2.05% (Table 3.16).

As a result of the increasing trend of the values of seasonal maximum and minimum temperatures, a decreasing trend will exist in maize productivity. The lowest value of future maize productivity will exist in 2080–2099. The projected values of maize productivity will be lower than the baseline value, with a mean value of 5.38 ton/ha and a range between minimum and maximum values of 0.89 ton/ha. The lowest percentage of reduction between baseline and projected productivity

Table 3.16 Projected climate elements and maize productivity in the future under SSP5-8.5 future climate scenario in the second agro-climatic zone

	T_{max} (°C)	T_{min} (°C)	RH (%)	Projected maize productivity (ton/ha)	Percentage of reduction (%)
2020–2039	35.42	21.21	50.97	5.80	35
2040–2059	36.65	22.35	50.62	5.55	38
2060–2079	38.01	23.82	49.83	5.26	41
2080–2099	39.63	25.54	48.92	4.91	45
Mean	37.43	23.23	50.08	5.38	40
Range	4.21	4.33	2.05	0.89	10

T_{max} = maximum temperature (°C), T_{min} = minimum temperature (°C), RH = relative humidity (%)
Baseline maize productivity = 8.96 ton/ha

will occur in 2020–2039 and the highest percentage will occur in 2080–2099, with an average of 40%, and a range between minimum and maximum values of 10% (Table 3.16).

Previous research by Taha (2012) showed a decline in maize productivity in 2030 by 38% in the second agro-climatic zone.

The Third Agro-climatic Zone

A similar trend observed in the first and second agro-climatic zone will be found in the third agro-climatic zone, where the values of seasonal maximum and minimum temperatures will gradually increase until 2080–2099 and the values of seasonal relative humidity will gradually decline. The mean values of the seasonal maximum and minimum temperatures will reach 40.75 and 24.52 °C, respectively with a range between minimum and maximum values will reach 4.63 and 4.80 °C, respectively. Nevertheless, the mean value of seasonal relative humidity will be 31.83% with a range between minimum and maximum values of 1.34% (Table 3.17).

The lowest value of future maize productivity will exist in 2080–2099, as a result of negative value of intercept of the regression equation, the high value of seasonal minimum temperature and its negative regression coefficient as presented in Table 3.8 for the third agro-climatic zone. The lowest value of percentage of reduction between baseline and projected maize productivity will exist in 2020–2039. High reduction in the projected maize productivity and the percentage of reduction will exist, with a mean value of 4.00 ton/ha and 54%, respectively and a range between minimum and maximum values of 2.33 ton/ha and 27%, respectively (Table 3.17).

In the third agro-climatic zone, simulation study by Ouda et al. (2009) projected a reduction in maize productivity in 2030 by 49%. Whereas, Moneir (2012) projected a decline in maize productivity by 51% in 2030 in the third agro-climatic zone.

Table 3.17 Projected climate elements and maize productivity in the future under SSP5-8.5 future climate scenario in the third agro-climatic zone

	T_{max} (°C)	T_{min} (°C)	RH (%)	Projected maize productivity (ton/ha)	Percentage of reduction (%)
2020–2039	38.48	22.23	32.60	5.10	42
2040–2059	39.98	23.61	32.01	4.53	48
2060–2079	41.44	25.21	31.26	3.60	59
2080–2099	43.11	27.02	31.44	2.77	68
Mean	40.75	24.52	31.83	4.00	54
Range	4.63	4.80	1.34	2.33	27

T_{max} = maximum temperature (°C), T_{min} = minimum temperature (°C), RH = relative humidity (%)
Baseline maize productivity = 8.77 ton/ha

The Fourth Agro-climatic Zone

In the fourth agro-climatic zone, the values of seasonal maximum and minimum temperatures and relative humidity followed the same pattern as in the previous agro-climatic zones. The mean values of the seasonal maximum and minimum temperatures and relative humidity will reach 41.47 °C, 24.67 °C and 27.79%, respectively with a range between minimum and maximum values will reach 3.94 °C, 4.81°C, and 2.22% respectively (Table 3.18).

The lowest value of future maize productivity will exist in 2080–2099. This is due to the reduction occurs in seasonal relative humidity and its negative regression coefficient in the prediction equation presented in Table 3.8 in the fourth agro-climatic zone. The projected values of maize productivity will be lower than the baseline value, with a mean value of 4.25 ton/ha and a range between minimum and maximum values of 0.10 ton/ha. The percentage of reduction between baseline and projected

Table 3.18 Projected climate elements and maize productivity in the future under SSP5-8.5 future climate scenario in the fourth agro-climatic zone

	T_{max} (°C)	T_{min} (°C)	RH (%)	Projected maize productivity (ton/ha)	Percentage of reduction (%)
2020–2039	39.97	22.46	28.80	4.18	46
2040–2059	40.80	23.83	28.34	4.27	45
2060–2079	42.28	25.49	27.43	4.28	44
2080–2099	43.91	27.27	26.58	4.26	45
Mean	41.74	24.76	27.79	4.25	45
Range	3.94	4.81	2.22	0.10	2

T_{max} = maximum temperature (°C), T_{min} = minimum temperature (°C), RH = relative humidity (%)
Baseline maize productivity = 7.70 ton/ha

Table 3.19 Projected climate elements and maize productivity in the future under SSP5-8.5 future climate scenario in the fifth agro-climatic zone

	T_{max} (°C)	T_{min} (°C)	RH (%)	Projected maize productivity (ton/ha)	Percentage of reduction (%)
2020–2039	41.85	24.94	21.32	2.55	52
2040–2059	43.22	26.35	20.80	1.98	63
2060–2079	44.73	27.99	20.30	1.29	76
2080–2099	46.35	28.74	20.12	1.12	79
Mean	44.04	27.01	20.63	1.74	68
Range	4.50	3.80	1.21	1.43	27

T_{max} = maximum temperature (°C), T_{min} = minimum temperature (°C), RH = relative humidity (%)
Baseline maize productivity = 5.37 ton/ha

productivity will be 45%, and a range between minimum and maximum values of 2% (Table 3.18).

The Fifth Agro-climatic Zone

The values of seasonal maximum and minimum temperatures and relative humidity in the fifth agro-climatic zone followed the same pattern as in the previous agro-climatic zones. The values of seasonal maximum and minimum temperatures will gradually rise until 2080–2099 with mean values of 44.04 and 27.01 °C, respectively and a range between minimum and maximum values will reach 4.50 and 3.80 °C, respectively. Whereas, the seasonal values of relative humidity will be gradually decline until 2080–2099. The mean value of seasonal relative humidity will be 20.63% with a range between minimum and maximum values will be 1.21% (Table 3.19).

The highly increasing trend in maximum and minimum temperature clearly reflected on maize productivity and cause a decline in it, where the lowest value of future maize productivity will exist in 2080–2099. The projected values of maize productivity will be lower than the baseline value, with a mean value of 1.74 ton/ha and a range between minimum and maximum values of 1.43 ton/ha. The lowest percentage of reduction between baseline and projected productivity will occur in 2080–2099, with an average of 68%, and a range between minimum and maximum values of 27% (Table 3.19).

The Country of Egypt

The values of seasonal maximum and minimum temperatures will gradually rise until 2080–2099 as an average over all Egypt. The mean values of the seasonal

Table 3.20 Projected climate elements and maize productivity in the future under SSP5-8.5 future climate scenario in the country of Egypt

	T_{max} (°C)	T_{min} (°C)	RH (%)	Projected maize productivity (ton/ha)	Percentage of reduction (%)
2020–2039	39.12	23.73	29.13	5.22	34
2040–2059	40.50	25.09	28.71	4.92	38
2060–2079	41.94	26.64	28.05	4.62	42
2080–2099	43.61	28.40	27.63	4.23	47
Mean	41.29	25.96	28.38	4.75	40
Range	4.49	4.67	1.50	0.30	12

T_{max} = maximum temperature (°C), T_{min} = minimum temperature (°C), RH = relative humidity (%)
Baseline maize productivity = 7.94 ton/ha

maximum and minimum temperatures will reach 41.49 and 25.96 °C, respectively with a range between minimum and maximum values will reach 4.49 and 4.67 °C, respectively. Whereas, the seasonal values of relative humidity will be gradually decline until 2080–2099 with a mean value of 28.38% and a range between minimum and maximum values will be 1.50% (Table 3.20).

The lowest value of future maize productivity will exist in 2080–2099 as a result of the increasing trend in the seasonal maximum and minimum temperatures and decreasing trend in the seasonal relative humidity. The projected values of maize productivity will be lower than the baseline value, with a mean value of 4.75 ton/ha and a range between minimum and maximum values of 0.30 ton/ha. The lowest percentage of reduction between baseline and projected productivity will occur in 2080–2099, with an average of 40%, and a range between minimum and maximum values of 12% (Table 3.20).

Intercropping Legume Crops with Maize to Overcome Climate Variability and Change

Over the past century, the average surface temperature and frequent heatwaves have been steadily rising, especially during the summer season, which affecting the yield potential of most food crops. Hence, diversification in cropping systems is essential to ensure high yield potential. Legume-cereal intercropping plays a vital role in the existence of food production, particularly in situations with limited water resources (Dahmardeh et al., 2009), it can enhance the efficiency of capturing and utilizing solar energy, resulting in higher yields compared to sole cropping (Chen et al., 2020). Furthermore, legume-cereal intercropping can help in altering the abiotic and biotic characteristics of agro-ecosystems through incorporating different types of crops (Lin, 2011). The efficiency of legumes in fixing atmospheric nitrogen through symbiosis with rhizobia makes them mainly important in low-input agriculture (Bijarnia et al., 2021). It has been reported that soil moisture conservation in intercropping systems is attained as a result of early ground cover by the legume crop, which help in reduces soil water evaporation (Muoni et al., 2020). Furthermore, progressive differences in peak demand for nutrients and water at critical growth stages, such as grain filling reduces competition for resources between the component crops (Trail et al., 2016).

Under the prevailing conditions of global warming, significant impacts will be observed not only on crops productivity but also on agricultural insect pests, where temperature is probably the most important environmental factor affecting insect behavior, distribution, development and reproduction (Kocmánková et al., 2010). Climate change provide opportunities for insect pests to establish and spread in new geographic regions and shift from one region to another (Bale et al., 2002). Implementing intercropping systems can play an important role in reduce insects' infestation in one of the intercrops, as well as smother weeds (Trail et al., 2016). Legume-cereal intercropping systems could reduce nitrogen fertilizer by 26% (Jensen et al., 2020), which could shrink the carbon footprint of agriculture (Huss et al., 2022). Intercropping two crops on the same piece of land alters the condition and environmental quality of the host plant, which affects the ability of the insect pest to identify and recognize it (Maitra et al., 2019).

Several intercropping systems of legume crops (C3 crops) with maize (a C4 crop) were implemented globally, namely cowpea intercropped with maize, soybean intercropped with maize and peanut intercropped with maize. Intercropping a C3 crop with a C4 crop will have different type of CO_2 acquisition and fixation. Li et al. (2012) indicated that C4 have more efficient use of solar energy, cause an increase in its biomass with higher light intensity. Furthermore, the increase of both biomass and CO_2 fixation with higher CO_2 concentration in C4 are faster than that in C3, which reflect more efficient use of CO_2 in C4 plants (Wang et al., 2012). Thus, in this way, more quantity of greenhouse gas (CO_2) is used in photosynthesis and biomass production under intercropping of C3 and C4 plants. In addition, Wu et al. (2012) stated that the taller species in an intercropping system, such as maize, could change the light and heat environment of a shorter shade-tolerant species, such as cowpea.

Cowpea is adapted to drought stress under which many crops fail to grow well (Kyei-Boahen et al., 2017). It is a promiscuous grain legume that forms symbiosis with a wide range of indigenous strains of Rhizobia bacteria, forming the basis for biological nitrogen fixation (Smith et al., 2016). Hamd-Alla et al. (2014) reported that cowpea intercropped with maize resulted in an increase in maize yield by 10%. This increase in maize yield is attributed to a reduction in maize associated weeds (Zohry, 2005) and minimizing the number of predator's infestation in maize (Hamd-Alla et al., 2014). Furthermore, Dube et al. (2014) indicated that the gains in maize/cowpea intercrops may be attributed to better moisture conservation in intercrops as compared with sole crops.

Soybean intercropped with maize was found to play an important role in reducing pests of both crops. Chabi-Olaye et al. (2005) indicated that larval density in maize was reduced from 48.1 to 21.3% under intercropping, compared to maize mono cultivation. Furthermore, maize mono cultivation resulted in more tunneled stems and damaged cobs, and consequently lower maize yield. Bilalis et al. (2010) stated that, when compared to mono crop, weed density was lower in maize-soybean intercrops, were weeds had less access to light under this system. Whereas, Gao et al. (2014) indicated that intercropping soybean with maize significantly lowered disease incidence and the severity index of soybean red crown rot.

The system of peanut intercropped with maize has many benefits. This intercropping system not only improves crop yield, but also improves the efficiency of water and fertilizer utilization (Wang et al., 2020). Moreover, it reduces the competition for soil nutrients, increases beneficial soil microorganism numbers and diversity, as well as increases the nitrogen content of intercropped maize (Zhao et al., 2022). It reduces pathogenic and poisonous microorganisms (Chen et al., 2018), in addition to helping in reducing carbon emissions (Jiao et al., 2021). Chen et al. (2018) stated that maize root exudates promote the expression of chalcone-flavanone isomerase and the synthesis of flavonoids in peanut roots, which increase nodulation and nitrogen fixation.

Conclusion

Climate variability and change will have a harmful effect on maize productivity in the agro-climatic zones of Egypt. Our results showed that the assessment of the effect of climate variability on the level of the agro-climatic zones of Egypt on maize productivity gave more accurate results than on Egypt as a whole. Our results also presented the expected future climate of Egypt under the two future climate scenarios, which could help other researchers in their climatological studies. Using the regression analysis enable to develop prediction equation for maize productivity in the future under both future scenarios. One way to reduce the effect of climate variability and change on maize productivity is to intercrop it with legume crops, which can help in reducing insects' infestation and control weeds; reduce CO_2 emission from the soil and reduce carbon footprint of agriculture, better moisture conservation and use; and increasing soil fertility through fixing atmospheric nitrogen.

References

Adhikari, U., Nejadhashemi, A. P., & Woznicki, S. A. (2015). Climate change and eastern Africa: A review of impact on major crops. *Food and Energy Security, 4*, 110–132.

Bale, J. S., Masters, G. J., Hodkinson, I. D., Awmack, C., Bezemer, T. M., Brown, V. K., Butterfield, J., Buse, A., Coulson, J. C., Farrar, J., et al. (2002). Herbivory in global climate change research: Direct effects of rising temperature on insect herbivores. *Global Change Biology, 8*, 1–16. https://doi.org/10.1046/j.1365-2486.2002.00451.x

Banziger, M., & Diallo, A. O. (2004). Progress in developing drought and N stress tolerant maize cultivars for Eastern and Southern Africa. In *Integrated approaches to higher maize productivity in the new millennium. Proceedings of the 7th Eastern and Southern Africa regional maize conference, CIMMYT/KARI*; Nairobi, Kenya, pp. 189–194.

Bates, B. C., Kundzewicz, Z. W., Wu, S., & Palutikof, J. P. (Eds.) (2008). Climate change and water. Technical Paper of the Intergovernmental Panel on Climate Change, IPCC Secretariat, Geneva, 210 pp.

Bijarnia, A., et al. (2021). Effect of nitrogen and phosphorus fertilization along with foliar spray of calcium chloride and potassium nitrate on performance of cowpea (Vigna unguiculata) + baby corn (Zea mays) intercropping systems. *Indian Journal of Agronomy, 66*, 355–360.

Bilalis, D., Papastylianou, P., Konstantas, A., Patsiali, S., Karkanis, A., & Efthimiadou, A. (2010). Weed-suppressive effects of maize-legume intercropping in organic farming. *International Journal of Pest Management, 56*(2), 173–181.

Chabi-Olaye, A., Nolte, C., Schulthess, F., & Borgemeister, C. (2005). Relationships of intercropped maize, stem borer damage to maize yield and land-use efficiency in the humid forest of Cameroon. *Bulletin of Entomological Research, 95*(5), 417.

Chen, J., Arafat, Y., Wu, L., Xiao, Z., Li, Q., Khan, M. A., et al. (2018). Shifts in soil microbial community, soil enzymes and crop yield under peanut/maize intercropping with reduced nitrogen levels. *Applied Soil Ecology, 124*, 327–334.

Chen, J., Xu, W., Burke, J. J., & Xin, Z. (2010). Role of phosphatidic acid in high temperature tolerance in maize. *Crop Science, 50*, 2506–2515.

Chen, X., Sun, N., Gu, Y., Zheng, H., Li, J., Wu, C., & Wang, Z. (2020). Photosynthetic and chlorophyll fluorescence responses in maize and soybean strip intercropping system. *International Journal of Agriculture and Biology, 24*, 799–811. https://doi.org/10.17957/IJAB/15.1503

Dahmardeh, M., Ghanbari, A., Syasar, B., & Ramrodi, M. (2009). Intercropping maize (Zea mays L.) and cowpea (Vigna unguiculata L.) as a whole-crop forage. *Journal of Food, Agriculture, Environment, 7*, 505–509.

Draper, N. R., & Smith, H. (1987). *Applied regression analysis* (pp. 397–402). John Wiley and Sons, Inc.

Dube, E. D. N., Madanzi, T., Kapenzi, A., & Masvaya, E. (2014). Root length density in maize/cowpea intercropping under a basin tillage system in a semi-arid area of Zimbabwe. *American Journal of Plant Sciences, 5*, 1499–1507. https://doi.org/10.4236/ajps.2014.511165

Erenstein, O., Jaleta, M., Sonder, K., et al. (2022). Global maize production, consumption and trade: Trends and R&D implications. *Food Security, 14*, 1295–1319. https://doi.org/10.1007/s12571-022-01288-7

Eyring, V., Bony, S., Meehl, G. A., Senior, C. A., Stevens, B., Stouffer, R. J., & Taylor, K. E. (2016). Overview of the coupled model intercomparison Project Phase 6 (CMIP6) experimental design and organization. *Geoscientific Model Development, 9*(5), 1937–1958.

Farré, I., & Faci, J. M. (2006). Comparative response of maize (Zea mays L.) and sorghum (Sorghum bicolor L. Moench) to deficit irrigation in a Mediterranean environment. *Agricultural Water Management, 83*, 135–143.

Gamal, G., Samak, M., & Shahba, M. (2021). The possible impacts of different global warming levels on major crops in Egypt. *Atmosphere, 12*, 1589. https://doi.org/10.3390/atmos12121589

Gao, X., Wu, M., Xu, R., Wang, X., Pan, R., Kim, H.-J., et al. (2014). Root interactions in a maize/soybean intercropping system control soybean soil-borne disease, red crown rot. *PLoS ONE, 9*(5), e95031.

Gomez, K. A., & Gomez, A. A. (1984). *Statistical procedures for agricultural research* (2nd ed.). John Willey and Sons.

Hamd-Alla, W. A., Shalaby, E. M., Dawood, R. A., & Zohry, A. A. (2014). Effect of cowpea (Vigan sinensis L.) with maize (Zea mays L.) intercropping on yield and its components. *International Scholarly and Scientific Research and Innovation, 8*(11), 1170–1176.

Hoffmann, U. (2013). Section B: Agriculture: a key driver and a major victim of global warming, in: Lead Article, in: Chapter 1, in Hoffmann, pp. 3–5.

Hund, A., Fracheboud, Y., Soldati, A., & Stamp, P. (2008). Cold tolerance of maize seedlings as determined by root morphology and photosynthetic traits. *European Journal of Agronomy, 28*, 178–185. https://doi.org/10.1016/j.eja.2007.07.003

Huss, C. P., Holmes, K. D., & Blubaugh, C. K. (2022). Benefits and risks of intercropping for crop resilience and pest management. *Journal of Economic Entomology, 115*(5), 1350–1362. https://doi.org/10.1093/jee/toac045

IPCC. (2014). Climate change 2014: Impacts, adaptation, and vulnerability. *Part A: Global and sectoral aspects. Summary for policymakers* (pp. 1–32). Contribution of Working Group II to the Fifth Assessment Report of the Intergovernmental Panel on Climate Change (IPCC). Cambridge University Press.

Jamieson, P. D., Porter, J. R., Goudriaan, J., Ritchie, J. T., Keulen, H., & Stol, W. (1998). A comparison of the models AFRCWHEAT2, CERES-Wheat, Sirius, SUCROS2 and SWHEAT with measurements from wheat grown under drought. *Field Crops Research, 55*, 23–44.

Jensen, E. S., Carlsson, G., & Hauggaard-Nielsen, H. (2020). Intercropping of grain legumes and 748 cereals improves the use of soil N resources and reduces the requirement for synthetic 749 fertilizer N: A global-scale analysis. *Agronomy for Sustainable Development, 40*, 750. https://doi.org/10.1007/s13593-020-0607-x

Jiao, N., Wang, F., Ma, C., Zhang, F., & Jensen, E. S. (2021). Interspecific interactions of iron and nitrogen use in peanut (Arachis hypogaea L.)-maize (Zea mays L.) intercropping on a calcareous soil. *European Journal of Agronomy, 128*.

Kim, Y. H., Min, S. K., Zhang, X. B., Sillmann, J., & Sandstad, M. (2020). Evaluation of the CMIP6 multi-model ensemble for climate extreme indices. *Weather and Climate Extremes, 29*. Article 100269.

Kocmánková, E., Trnka, M., Juroch, J., Dubrovský, M., Semerádová, D., Možný, M., Žalud, Z., Pokorný, R., & Lebeda, A. (2010). Impact of climate change on the occurrence and activity of harmful organisms. *Plant Protection Science, 45*, S48–S52. https://doi.org/10.17221/2835-PPS

Kyei-Boahen, S., Savala, C. E. N., Chikoye, D., Abaidoo, R., & Kyei-boahen, S. (2017). Growth 762 and yield responses of cowpea to inoculation and phosphorus fertilization in different 763 environments. *Frontiers in Plant Science, 8*, 1–13. https://doi.org/10.3389/fpls.2017.00646

Li, C. S., Tang, Y. L., Wu, C., & Huang, G. (2012). Effect of sowing patterns on growth, development and yield formation of wheat in rice stubble land. *Transactions of CSAE, 28*, 36–43.

Lin, B. B. (2011). Resilience in agriculture through crop diversification: Adaptive management for environmental change. *BioScience, 61*, 183–193. https://doi.org/10.1525/bio.2011.61.3.4

Lobell, D. B., Bänziger, M., Magorokosho, C., & Vivek, B. (2011). Nonlinear heat effects on African maize as evidenced by historical yield trials. *Nature Climate Change, 1*, 42–45.

Lobell, D. B., Hammer, G. L., McLean, G., Messina, C., Roberts, M. J., & Schlenker, W. (2013). The critical role of extreme heat for maize production in the United States. *Nature Climate Change, 3*, 497–501.

Maitra, S., Palai, J. B., Manasa, P., & Kumar, D. P. (2019). Potential of intercropping system in sustaining crop productivity. *International Journal of Agriculture, Environment and Biotechnology, 12*(1), 39–45.

Moneir, I. (2012). *Simulation of the effect of adaptation strategies on improving yield of some crops grown under expected climate change conditions* [MSc Thesis. Institute of Environmental Research Studies, Ain Shams University, Egypt].

Morsy, M. (2015) *Use of regional climate and crop simulation models to predict wheat and maize productivity and their adaptation under climate change* [PhD thesis, Faculty of Science Al-Azhar University].

Morsy, M., Sayad, T., & Ouda, S. (2017). Present and future water requirements for crops. In *Future of food gaps in Egypt: Obstacles and opportunities*. Springer Publishing House. ISBN 978-3-319-46942-3.

Motha, R. P. (2011). The impact of extreme weather events on agriculture in the United States. In *Challenges and opportunities in agrometeorology*. https://doi.org/10.1007/978-3-642-19360-6_30

Msowoya, K., Madani, K., Davtalab, R., Mirchi, A., & Lund, J. R. (2016). Climate change impacts on maize production in the warm heart of Africa. *Water Resources Management, 30*, 5299–5312. https://doi.org/10.1007/s11269-016-1487-3

Muoni, T., Koomson, E., Öborn, I., Marohn, C., Watson, C. A., Bergkvist, G., Barnes, A., Cadisch, G., & Duncan, A. (2020). Reducing soil erosion in smallholder farming systems in East Africa through the introduction of different crop types. *Experimental Agriculture, 56*, 183–195. https://doi.org/10.1017/S0014479719000280

Ouda, S., Abdrabbo, M., & Noreldin, T. (2012). Effect of changing sowing dates and irrigation scheduling on maize yield grown under climate change conditions. In *10th International conference of Egyptian soil science society (ESSS) and 4th international conference of water requirements & metrology department, 5–7 November, Ameria, Egypt (Cl 07).*

Ouda, S., & Noreldin, T. (2017). Evapotranspiration data to determine agro- climatic zones in Egypt. *Journal of Water and Land Development, 32*(I–III), 79–86.

Ouda, S. A., Khalil, F. A., & Yousef, H. (2009). Using adaptation strategies to increase water use efficiency for maize under climate change conditions. In *Proceeding of 13th international conference on water technology, Egypt.*

Poole, N., Donovan, J., & Erenstein, O. (2021). Agri-nutrition research: Revisiting the contribution of maize and wheat to human nutrition and health. *Food Policy, 100*, 101976. https://doi.org/10.1016/j.foodpol.2020.101976

Rinaldi, M. (2009). A simulation approach to investigate options for mitigation and crop adaption to climate change in Mediterranean area. In *2009 IOP conference series: Earth environmental science.* https://doi.org/10.1088/1755-1307/6/7/372038

Saleh, S. M. M. (2007). *Study of the effect of expected climatic changes on potato productivity in Egypt* [PhD thesis, Department of Horticulture, Faculty of Agriculture, Ain Shams University].

Shiferaw, B., Prasanna, B., Hellin, J., & Banziger, M. (2011). Crops that feed the world 6. Past successes and future challenges to the role played by maize in global food security. *Food Security, 3*, 307–327. https://doi.org/10.1007/s12571-011-0140-5

Smith, A., Snapp, S., Dimes, J., Gwenambira, C., & Chikowo, R. (2016). Doubled-up legume 898 rotations improve soil fertility and maintain productivity under variable conditions in 899 maize-based cropping systems in Malawi. *Agricultural Systems, 145*(139–149), 900. https://doi.org/10.1016/j.agsy.2016.03.008

Taha, A. (2012). *Effect of climate change on maize and wheat grown under fertigation treatments in newly reclaimed soil* [PhD Thesis, Tanta University, Egypt].

Teixeira, E. I., Fischer, G., Velthuizen, H. V., Walter, C., & Ewert, F. (2013). Global hot spots of heat stress on agricultural crops due to climate change. *Agricultural and Forest Meteorology, 170*, 206–215. https://doi.org/10.1016/j.agrformet.2011.09.002

Tesfaye, K., Gbegbelegbe, S., Cairns, J. E., Shiferaw, B., Prasanna, B. M., Sonder, K., et al. (2015). Maize systems under climate change in sub-Saharan Africa: Potential impacts on production and food security. *International Journal of Climate Change Strategies and Management*, 247–271.

Trail, P., Abaye, O., Thomason, W. E., Thompson, T. L., Gueye, F., Diedhiou, I., Diatta, M. B., & Faye, A. (2016). Evaluating intercropping (living cover) and mulching (desiccated cover) practices for increasing millet yields in Senegal. *Agronomy Journal, 108*, 1742–1752. https://doi.org/10.2134/agronj2015.0422

Wang, L., Guo, L., Li, Y., Wang, Z., et al. (2012). Systematic comparison of C3 and C4 plants based on metabolic network analysis chuanli BMC. *Systems Biology, 6*(Suppl 2), S9. http://www.biomedcentral.com/1752-0509/6/S2/S9

Wang, R., Sun, Z., Zhang, L., Yang, N., Feng, L., Bai, W., et al. (2020). Border-row proportion determines strength of interspecific interactions and crop yields in maize/peanut strip intercropping. *Field Crop Research,* 253.

Willmott, C. J. (1981). On the validation of models. *Physical Geography, 2*, 184–194.

World Bank group (2021) Climate Change Knowledge Portal (CCKP) User Manual. World Bank Group. 18 pages.

Wu, W., Fan, L., Li, M. F., Liu, H. B., & Li, Y. Q. (2012). Sensitivity analysis of crop growth models to multi-temporal scale solar radiation. *Transactions of the Chinese Society of Agricultural Engineering, 28*, 123–128.

Zhao, C., Liu, B., Piao, S., Wang, X., Lobell, D. B., Huang, Y., Huang, M., Yao, Y., Bassu, S., Ciais, P., Durand, J.-L., Elliott, J., Ewert, F., Janssens, I. A., Li, T., Lin, E., Liu, Q., Martre, P., Müller, C., Peng, S., …, Asseng, S. (2017). Temperature increase reduces global yields of major crops in four independent estimates. *PNAS, 114*(35), 9326–9331. https://doi.org/10.1073/pnas.1701762114

Zhao, F., Zhang, D., Zhao, Y., Wang, W., Yang, H., Tai, F., et al. (2016). The difference of physiological and proteomic changes in maize leaves adaptation to drought, heat, and combined both stresses. *Frontiers in Plant Science, 7*, 1471. https://doi.org/10.3389/fpls.2016.01471

Zhao, X., Dong, Q., Han, Y., et al. (2022). Maize/peanut intercropping improves nutrient uptake of side-row maize and system microbial community diversity. *BMC Microbiology, 22*, 14. https://doi.org/10.1186/s12866-021-02425-6

Zohry, A. A. (2005). Effect of relaying cotton on some crops under bio-mineral N fertilization rates on yield and yield components. *Annals of Agricultural Science, 431*, 89–103.

Zohry, A. A., & Ouda, S. (2017). Increasing land and water productivities to reduce maize food gap. In *Future of food gaps in Egypt: Obstacles and opportunities*. Springer Publishing House. ISBN 978-3-319-46942-3

Chapter 4
Legumes Improve Wheat and Maize Productivity Grown in Different Cropping Systems Under Changing Climate

Abstract Diversification of agricultural cropping systems is known to be a significant strategy to accomplish the ecological intensification and enhancing sustainability. This chapter composed of two sections. In the first section, we tested the capability of SALTMED model to simulate multi-year crop sequences of maize and wheat, as well as simulation of the inclusion of legume crops, namely fahl clover and cowpea within the system of maize and wheat in Egypt. The SALTMED model was calibrated and validated using data of a two-year field experiment for three crop sequences for maize and wheat in sandy soil: CS1, which represent the traditional practice by the farmers (maize-fallow-wheat), and involved low water and fertilizer use efficiency. The other two crop sequences implemented irrigation scheduling and fertigation: CS2 (maize-fahl clover-wheat) and CS3 (cowpea intercropped with maize-fahl clover-wheat). The results indicated that the SALTMED model was successful in simulating the yield of the studied crop sequences, with high degree of accuracy. The model also simulated nitrogen uptake by the studied crops, as well as the amount of nitrogen leached in the end of each growing seasons. In the second section of this chapter, we evaluated the role of legumes in enhancing the resiliency of maize and wheat to climate change under the studied crop sequences using SALTMED model and CMIP6 mean projection of two future scenarios, namely SSP1-2.6 and SSP5-8.5. Under future climate scenarios, the model simulated the losses in the yield and the season length of the studied crops, where the highest yield losses will be found under CS1 and the lowest yield losses will be found under CS3. Nitrogen use efficiency and water productivity values of the studied crops will be the lowest under CS1 and highest under CS3. The above results proved that the integration of legume crops in maize and wheat cropping system can increase its resiliency and mitigate the harmful effect of climate change, as well as reducing yield losses.

Keywords SALTMED model · Nitrogen uptake and leaching · SSP1-2.6 · SSP5-8.5 · Nitrogen use efficiency · Water productivity

S. Ouda and A. E. Zohry, *Integration of Legume Crops with Cereal Crops Under Changing Climate*, https://doi.org/10.1007/978-3-031-68102-8_4

Introduction

Diversification of agricultural cropping systems is known to be a significant strategy to accomplish the ecological intensification and enhancing sustainability (Wezel et al., 2014). The sustainable method to meet future demands for food and protein is to include legume in the cropping system (Foyer et al., 2016). This method attains diversification of agricultural cropping systems and could replace agro-chemicals and fossil fuels with more ecological practices to increase food security (Bommarco et al., 2013). The benefits of using a legume crop as a rotation crop have been well documented. Nuruzzaman et al. (2005) stated that a legume crop helps in the introduction of atmospheric nitrogen into agricultural systems through symbiotic nitrogen fixation, which has the potential to enhances soil nitrogen for subsequent cereal crop (Ruisi et al., 2017). Inclusion of legumes in the cropping system can reduce the need to nitrogen fertilizers, thus lower the emission of greenhouse gases, such as carbon dioxide and nitrous oxide (Lemke et al., 2007). It also can improve soil health and structure, have an important role in the sequestration of carbon in soils (Lopez-Bellido et al., 2006). Legumes, not only, increase soil nitrogen content, but also enhance phosphorus availability (Pypers et al., 2007) and provide a favorable microbial community in the rhizosphere (Yusuf et al., 2009). Consequently, it positively affects the soil properties, including soil biological, chemical, and physical properties (Vasconcelos et al., 2020). Thus, the inclusion of legumes in the cropping systems to make use of its associated biological nitrogen fixation systems provides an economically feasible and environmentally sound method to reduce the external inputs, improving the soil nutrient content and, can be suggested as a sustainable method to increase food security (Mayer et al., 2003).

The extent of biological nitrogen fixation and its associated contribution differs across legume species, climatic conditions, cropping systems, and soil properties, as well as soil management practices (Stagnari et al., 2017), for example the biological nitrogen fixation resulted from Egyptian clover is 30–59 kg/ha and from cowpea is 36–75 kg/ha (Vasconcelos et al., 2020).

The mechanisms of contribution of legumes in a cropping system has two ways, namely direct nutrients transfer and non-direct nutrients transfer. According to Thilakarathna et al. (2016), the direct nutrients transfer is the transfer of nutrients from a nitrogen-fixing legume to another crop during the growth of the crop in an intercropping system and/or as remaining nitrogen for the advantage of a subsequent crop through decayed legume nodules, roots, root border cells, root caps, sloughed cells, and the epidermis. Ghosh et al. (2007) stated that cowpea cultivated as an intercrop with maize had a positive residual contribution on the yield of the following wheat crop.

The non-direct nutrients transfer can be visualized by the fact that legumes-based cropping systems improve several aspects of soil fertility, including soil organic carbon and humus content, nitrogen and phosphorous availability and uptake, and organic carbon and nitrogen, as well as releasing hydrogen gas as a by-product of biological nitrogen fixation, which promotes bacterial legume nodules' development in the rhizosphere (Stagnari et al., 2017).

Traditional agronomic field experiments have been and will be the reliable information sources to establish relationships between agricultural management practices and real-world observed measurements (Durr et al., 2016). However, as a result of increasing demands for data for decision-makers, the traditional agronomic research methods are becoming insufficient, because they are usually present results from certain seasons and it is site-specific (Patricia et al., 2013). In addition, these trials are usually expensive, time consuming and labor intensive (Choruma et al., 2019). Therefore, there is an urgent need for a tool/method to quickly and easily obtain research findings. Hence, the crop simulation models have been developed to predict crops yield cultivated under various management scenarios and time scales (Rötter et al., 2011). For instance, the use of crop simulation models offers an opportunity for exploring cultivar potential for new areas before establishing expensive and time-consuming field experiments (Bationo et al., 2012). Moreover, field trials with a high number of treatments could be pre-evaluated by simulation models (Steduto et al., 2009). Additionally, assessing the impacts of climate change/variability and management practices on crop productivity and environmental can be also evaluated by simulation models (Kadiyala et al., 2015).

SALTMED model (Ragab, 2002, 2015) is one of the important process-oriented crop simulation models. The model uses integrated crop–soil–water processes and have been proven to be an effective tool for water resources management in agriculture (Ragab, 2015). The model has been widely used for different crops irrigated with different irrigation and fertigation systems implemented in different parts of the world. It has been used in Egypt and Syria for tomato crop (Ragab et al., 2005a, 2005b), for sugarcane in Iran (Golabi et al., 2009), and for chickpea in Portugal (Silva et al., 2013). It has been also used for quinoa in Morocco (Fghire et al., 2015; Hirich et al., 2012), in Denmark (Razzaghi et al., 2011), and in Italy (Pulvento et al., 2013). It has been used for amaranth in Italy (Pulvento et al., 2015), and for legumes in Syria (Arslan et al., 2015). A detailed description of the SALTMED model have been provided by Ragab (2002, 2015) and Ragab et al., (2005a, 2005b).

Maize (a summer crop) and wheat (a winter crop) are two important crops in Egypt usually preceding each other. Such a practice usually causes low productivity of both crops, as a result of nutrients depletion from the soil. To solve this problem and enhance the productivity of both crops, inclusion of a legume crop in this crop sequence, or intercrop a legume crop with maize could be done.

Fahl clover is a good candidate to be planted after maize to precede wheat. Fahl clover is a variety of Egyptian clover, which has stem branching ability and is cut only once. It is characterized by rapid growth and large forage yield, and it can be cultivated as early winter season crop starting in September and grow until the beginning of November, where its growing season is between 60 and 70 days (Bakheit et al., 2016), thus it can be considered as an improved fallow. Improved fallow involves planting a beneficial legume crop during the fallow period to restore soil fertility and enhance subsequent crop production. The nutrients supply, such as nitrogen, is greater in improved fallow than in cropped land as the plants store nutrients from the air in the soil, and falls their leaf litter to improve the soil and preserve moisture (Lemage & Tsegaye, 2020). Awad et al. (2023) indicated that soil nitrogen, phosphorous and potassium were increased when fahl clover preceded wheat, and wheat yield was increased by 8%, compared to maize preceded wheat. Whereas, Hamada and Hamd-Alla (2019) indicated that fahl clover cultivation before wheat increased wheat yield by 6%, compared to fallow preceding wheat.

Cowpea is an annual fodder legume crop characterized by being a drought-tolerant grown for its nutritious fodder for livestock. It is a fast-growing leguminous fodder that can be grown as a sole crop, where it has the unique ability to fix nitrogen even in very poor soil (Singh, 2014). It is also characterized by being a shade-tolerance, therefore it is compatible as an intercrop with a number of cereals and root crops (Boukar et al., 2013). When intercropped with cereals, cowpea can improve soil fertility through compensate for the loss of soil nitrogen absorbed by cereals. It is also a good cover crop that can limits soil erosion (Singh et al., 2003). Zohry et al. (2017) indicated that intercropping cowpea with maize resulted in an increase in maize yield by 8%, compared to maize yield obtained under monoculture. Whereas, Awad et al. (2023) reported that when cowpea intercropped with maize, followed by fahl clover and then wheat, the highest percentage of increase in wheat yield was found, namely 20%, compared to the yield obtained from wheat monoculture. Thus, implementing such crop sequences of maize and wheat with one or two legume crops included, could increase maize and wheat productivities and soil nitrogen percentage (Ouda et al., 2020). Using SALTMED model to simulate the crop yield of such sequences can help in improving our understanding of the effect of inclusion of these legume crops on soil nitrogen content.

This chapter composed of two sections. In the first section, we tested the capability of SALTMED model to simulate multi-year crop sequences of maize and wheat, as well as simulation of the inclusion of legume crops, namely fahl clover and cowpea within the sequence of these two important cereal crops in Egypt. The model also has the ability to simulate nitrogen uptake and leaching, which could be an indicator of the improvement of the soil due to the inclusion of legume crops.

In the second section of this chapter, we evaluated the role of legumes in enhancing the resiliency of maize and wheat to climate change when grown under different cropping systems using SALTMED simulation model and CMIP6 mean projection of two future scenarios, namely SSP1-2.6 and SSP5-8.5 in two-time intervals, namely 2060–2079 and 2080–2099. The effect of climate change on season length of maize and wheat was also assessed, as well as its effect on nitrogen use efficiency and water productivity of the studied crop sequences.

Calibration and Validation of SALTMED Model

The Data of the Field Experiments

The data of a two-year field experiment was obtained from Ouda et al. (2020). In a complete blocks design with four replications, two-year field experiment was conducted at in the sandy soil of El-Minia Governorate (latitude = 28.05°, longitude = 30.44° and elevation above sea level = 40.0 m); Middle Egypt during the two growing seasons of 2016/17 and 2017/18 under sprinkler system.

Three crop sequences for maize and wheat were simulated and evaluated with respect to the resulted yield of both crops, nitrogen uptake and nitrogen leaching. The first crop sequence (CS1) is the traditional practice by the farmers in the surrounding areas, where maize is followed by fallow, then wheat. This sequence involved high applied irrigation amounts without scheduling, in addition to broadcasting fertilizer on the soil before irrigation, which resulted in low water and fertilizer use efficiency. Irrigation scheduling is a process used to determine the amount of water applied to the crop and the timing for application (Fernandez & Cuevas, 2010). The other two suggested crop sequences implemented irrigation scheduling and fertigation, namely maize, improved fallow (fahl clover) then wheat (CS2); and cowpea intercropped with maize, improved fallow (fahl clover) then wheat (CS3).

The values of monthly reference evapotranspiration (ETo) were calculated using the Penman–Monteith equation (Allen et al., 1998) (Fig. 4.1a). The applied irrigation water amounts (Fig. 4.1b) under standard conditions using a sprinkler system were determined by the following equation:

$$\text{ETc} = \text{ETo} \times K_c \tag{4.1}$$

where: ETc = crop evapotranspiration (mm/day), ETo = reference evapotranspiration (mm/day), and K_c = crop coefficient.

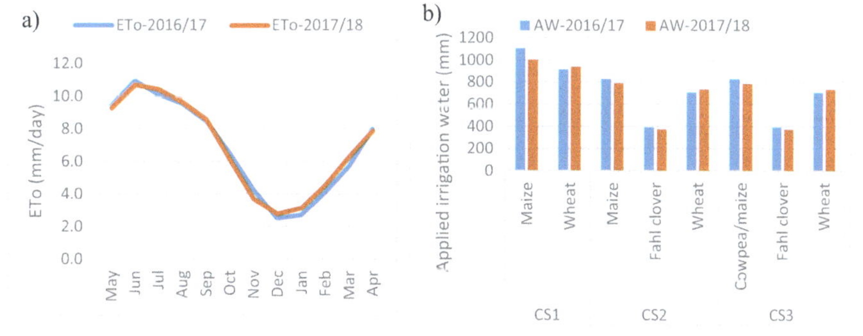

Fig. 4.1 Monthly values of ETo (**a**) and the amount of applied irrigation water to the studied crops (**b**) during the two growing seasons of the field experiment

It is clear from Fig. 4.1a that ETo values in 2016 were slightly higher in May and June and lower in July and August, compared to its values in 2017, which resulted in higher value of applied irrigation water to maize in the first season, compared to the second season (Fig. 4.1b). Similarly, ETo values were higher in September and October in 2017, compared to its value in 2016 (Fig. 4.1a), which resulted in higher applied irrigation water to fahl clover in 2017, compared to its value in 2016 (Fig. 4.1b). This situation was reversed in wheat, where ETo values were higher starting from December until March in 2017/18 (Fig. 4.1a) and that resulted in higher applied irrigation water in the second season, compared to the first season (Fig. 4.1b).

Table 4.1 presents the applied fertilizer amounts to the studied crops in both growing seasons. Nitrogen fertilizer as ammonium nitrate was applied in five doses to maize and wheat and in three doses to fahl clover. Cowpea did not receive any extra fertilizer or irrigation because it shared the applied amounts with maize under its intercropping system. Furthermore, calcium super phosphate (15.5% P_2O_5) and potassium sulphate (48.8% K_2O) were applied during land preparation.

Table 4.2 presents the yield of each crop in the studied crop sequences in both growing seasons. Comparing the yield of the studied crops in both growing seasons reveled a decrease in the yield of maize and wheat in CS1 in the second growing season, as a result of the depletion of soil nutrients after cultivation of two cereal crops, which implied its harmful effect on the soil. Hamd-Allah et al. (2014) indicated that low productivity of wheat was obtained when maize continuously preceded it. Whereas, a noticeable increase in the yield of all crops included in CS2 and CS3 was found. This increase could be attributed to using irrigation scheduling, which increase water productivity and the application of fertilizer through fertigation in these crop sequences, which minimizes fertilizer losses and increase fertilizer use efficiency (Taha & Ouda, 2016). Furthermore, it could be also attributed to the inclusion of

Table 4.1 Applied fertilizer amounts to the studied crops in both growing seasons

Crop sequence	Crop	N (kg/ha)	P_2O_5 (kg/ha)	K_2O (kg/ha)
CS1	Maize	360	58.6	74.4
	Wheat	288	74.4	58.6
CS2	Maize	360	58.6	74.4
	Fahl clover	72	37.2	0
	Wheat	288	74.4	58.6
CS3	Intercropped cowpea	0	0	0
	Intercropped maize	360	58.6	74.4
	Fahl clover	72	37.2	0
	Wheat	288	74.4	58.6

Source of data Ouda et al. (2020)
CS1 = maize, fallow, then wheat, CS2 = maize, improved fallow (fahl clover) then wheat, CS3 = cowpea intercropped with maize, improved fallow (fahl clover) then wheat

legume crops in these crop sequences. Sheha and et al. (2014) reported that cultivation of short season clover before wheat increased its yield.

Additionally, maize was benefited from the residual effect of the legume crop in the following season. Ouda et al. (2020) reported an increase in soil nitrogen after maize harvest by 88% in a two-year crop sequence, where fahl clover was cultivated between maize and wheat and by 113% after the harvest of cowpea intercropped with maize in a two-year crop sequence, where fahl clover was cultivated between maize intercropping system with cowpea and wheat. Furthermore, they stated that an increase by 75 and 88% in soil nitrogen was found after the harvest of wheat in these two crop sequences, respectively.

Furthermore, legume crops have the ability to facilitate the absorption of phosphorus and potassium in the soil by cereal crops, in addition to its role in providing nitrogen through nitrogen-fixing rhizobium (Ferguson et al., 2013), which positively affected the final yield. Bado et al. (2006) stated that nitrogen-fixing legumes supply nitrogen to the subsequent crops through fallen senescent leaves and below ground parts, leading to an increase in succeeding crop yield.

The high differences in the yield of the crops included in the three crop sequences in the two growing seasons make it a unique case to test the ability of the calibrated SALTMED to simulate these high differences in the yield and to visualize the benefits of legume crops inclusion. Thus, the field experimental data for both growing seasons were used as an input data in the SALTMED model. The data of the first growing season was used to calibrate the model and the data of the second growing season was used to validate the model. The calibration was done by fine tuning of "photosynthesis efficiency factor" to obtained reasonable values of the yield of the studied crops,

Table 4.2 The yield of the studied crops (ton/ha) under different crop sequences in both growing seasons

Crop sequence	Crop	First growing season	Second growing season	Percentage of change (%)
CS1	Maize	5.55	5.52	– 1
	Wheat	3.81	3.79	– 1
CS2	Maize	5.68	5.97	+ 5
	Fahl clover	4.41	5.23	+ 19
	Wheat	4.37	5.54	+ 27
CS3	Intercropped cowpea	3.04	3.83	+ 26
	Intercropped maize	5.74	6.24	+ 9
	Fahl clover	4.88	5.75	+ 18
	Wheat	4.72	6.16	+ 31

Source of data Ouda et al. (2020)
CS1 = maize, fallow, then wheat, CS2 = maize, improved fallow (fahl clover) then wheat, cowpea intercropped with maize, improved fallow (fahl clover) then wheat

whereas the validation was done by applying the calibrated values in the model to predict the yield of these crops.

The SALTMED model requires to input soil file, crop file, climate file, and nitrogen file. The soil of the experimental site is classified as sandy soil: 90.5% sand, 3.9% silt and 5.6% clay averaged over 60 cm depth. The bulk density of the soil was 1.6 mg/m, whereas field capacity, permanent wilting point, and available water were 9.9, 5.7, and 5.2% w/w. The above soil data was used to prepare the soil file, in addition to other required input data, which was imbedded in the model.

A weather file was prepared for each crop sequence starting from maize sowing date in the first growing season and ended with wheat harvest date in the second growing season. The weather data was obtained from the following website: https://power.larc.nasa.gov.

For the crop file required by the model, one file was prepared for each crop in each growing season. The input data required by SALTMED was obtained from measured field data (planting date, leaf area index, plant height, and the length of the growth stages) and are presented in Table 4.3. Other input data required by the model was obtained from the literature (root depth, Kc and Kcb; Allen et al. (1998)). Furthermore, we used the function "Fixed length growth stages" where harvest day is predetermined by the user.

Table 4.3 The input data obtained from the field experiments for the three crop sequences in both growing seasons

	First growing season			Second growing season		
	Leaf area index	Plant height (m)	Harvest index (%)	Leaf area index	Plant height (m)	Harvest index (%)
CS1						
Maize	2.80	1.90	0.39	2.80	1.80	0.39
Wheat	2.70	0.80	0.38	2.70	0.80	0.36
CS2						
Maize	3.00	1.90	0.39	2.80	1.90	0.42
Fahl clover	1.80	0.80	0.90	2.00	0.80	0.90
Wheat	3.40	0.80	0.40	4.20	1.10	0.46
CS3						
Intercropped maize	3.00	1.90	0.38	4.20	1.90	0.40
Intercropped cowpea	2.50	0.80	0.70	2.30	0.80	0.70
Cowpea/maize	3.00	1.90	0.39	4.20	1.90	0.40
Fahl clover	1.80	0.80	0.90	2.30	0.80	0.90
Wheat	4.00	0.90	0.41	4.70	1.10	0.47

CS1 = maize, fallow, and wheat, CS2 = maize, improved fallow (fahl clover) then wheat, CS3 = cowpea intercropped with maize, improved fallow (fahl clover) then wheat

Furthermore, the SALTMED model calculates the values of reference evapotranspiration (ETo) by several methods, among them is the FAO modified Penman–Monteith equation according to Allen et al. (1998), which was selected for this study. In each growing season, irrigation files under sprinkler system were prepared for each crop sequence. For CS1, fertigation was not applied because fertilizer was broadcasted on the soil, instead the applied nitrogen fertilizer was included in the nitrogen file, as required by the model's user guide. Whereas, for CS2 and CS3, fertigation was included in the irrigation file.

Regarding plant growth process for biomass calculation, SALTMED provides a detailed physiological imbedded factors in the "Crop growth tab" to be used in the calculation, or modified by the user. In CS3, cowpea was intercropped with maize and the SALTMED does not simulate intercropping systems. In this field experiment, the system of cowpea intercropped with maize shared the applied irrigation water and fertilizer with maize, in addition to cowpea was cultivated in the same date of maize and harvested the day before maize harvest. The amounts of water and fertilizer applied to that intercropping system produced the biomass of both crops. Therefore, first we simulated the grain yield of maize (100% of its recommended planting density) and the biological yield of cowpea (50% of its recommended planting density), separately. Then, we simulated the total measured biological yield of maize and cowpea together and calculated the percentage of cowpea contribution in the biological yield. The percentage of cowpea contribution in the biological yield of both crops in its intercropping system with maize was used in the function "Carbon Dioxide Impact Factor" included in the "Crop growth tab" to increase the production of biological yield by the value of the percentage of cowpea yield contribution in the biological yield. The "Carbon Dioxide Impact Factor" is originally used to tune the model against variation in the dry matter output as a result of potential future increases in global CO_2 levels. Nominally, this parameter has a value of unity. Thus, this factor helped us to simulate the biological yield of both cowpea and maize in its intercropping system, and its input values were 1.17 and 1.20 in the first and second growing season, respectively.

The "Rotation" tab was used to run the model for the crops included in each crop sequence in both growing seasons in the same time to produce the result of the whole crop sequence instead of each individual crop.

The Goodness of Fit Test

The goodness of fit between the measured yield values in each crop sequence and the predicted values by the SALTMED model was tested by coefficient of determination (R^2), root mean square error (RMSE) and Willmott index of agreements (*d-stat*) as followed.

1. **Coefficient of Determination (*R^2*)**

The R^2 value is calculated from the total sum of squares, more precisely, it is the sum of the squared deviations of the original data from the mean. R^2 takes on values between 0 and 1.0, a value of 1.0 indicates a perfect match, and value of 0 indicates no agreement at all. R^2 also shows how many points fall on the regression line (Jamieson et al., 1998).

$$R^2 = 1 - \frac{\sum_{i=1}^{n}(O_i - S_i)^2}{\sum_{i=1}^{n}\left(O_i - \overline{O}\right)^2} \tag{4.2}$$

where O_i and S_i represent the observed and predicted values, respectively, n represents the number of observed and predicted values used in comparison and $\overline{O}$ is the observed average.

2. **Root Means Square Error (RMSE)**

It is the standard deviation of the residuals (prediction errors). Residuals are a measure of how far from the regression line data points are. It is a frequently used to measure the difference between values predicted by a model and those actually observed from an experiment that is being modeled (Jamieson et al., 1998).

$$\text{RMSE} = \sqrt{\left(\frac{\sum_{i=1}^{n}(S_i - O_i)^2}{n}\right)} \tag{4.3}$$

3. **Willmott Index of Agreement (*d-stat*)**

It is the standardized measure of the degree of model prediction error and varies between 0 and 1.0. A value of 1.0 indicates a perfect match, and value of 0 indicates no agreement at all (Willmott, 1981).

$$d\text{-}stat = 1 - \frac{\sum_{i=1}^{n}(O_i - S_i)^2}{\sum_{i=1}^{n}\left[\left(\left|\left(S_i - \overline{O}\right)\right| + \left|\left(O_i - \overline{O}\right)\right|\right)^2\right]} \tag{4.4}$$

Calibration and Validation of SALTMED Model

Table 4.4 present the results of the goodness of fit test for both calibrated and validated values of the studied crops in the three crop sequences in both growing seasons. The result in the table proved that SALTMED model was able to simulate the growth environment of the studied crops in the field experiments with high level of accuracy as it was demonstrated by high value of R^2, and *d-stat*, as well as low value of RMSE in both calibrated and validated growing seasons.

Table 4.4 Measured versus predicted yield (ton/ha) of the studied crops in each crop sequence in both growing seasons, and percentage of yield reduction (PR%)

	First growing season (calibrated)			Second growing season (validated)		
	Measured	Predicted	PR (%)	Measured	Predicted	PR (%)
CS1						
Maize	5.55	5.35	4	5.52	5.20	6
Wheat	3.81	3.66	4	3.79	3.59	5
CS2						
Maize	5.68	5.46	4	5.97	5.60	6
Fahl clover	4.41	4.39	0.5	5.23	4.80	8
Wheat	4.37	4.19	4	5.54	5.28	5
CS3						
Intercropped maize	5.74	5.52	4	6.24	6.18	1
Intercropped cowpea	3.04	2.39	4	3.83	3.73	3
Cowpea/maize	17.76	17.40	2	18.69	18.49	1
Fahl clover	4.88	4.88	0	5.75	5.52	4
Wheat	4.72	4.51	4	6.16	5.60	9
R^2	0.9997			0.9988		
RMSE (ton/ha)	0.2045			0.3260		
d-stat	0.9994			0.9985		

CS1 = maize, fallow, then wheat, CS2 = maize, improved fallow (fahl clover) then wheat, CS3 = cowpea intercropped with maize, improved fallow (fahl clover) then wheat.

A significant positive relationship ($P < 0.01$) between the measured yield of the studied crops and its corresponded simulated values by the SALTMED model was found for each of the studied growing season (Fig. 4.2).

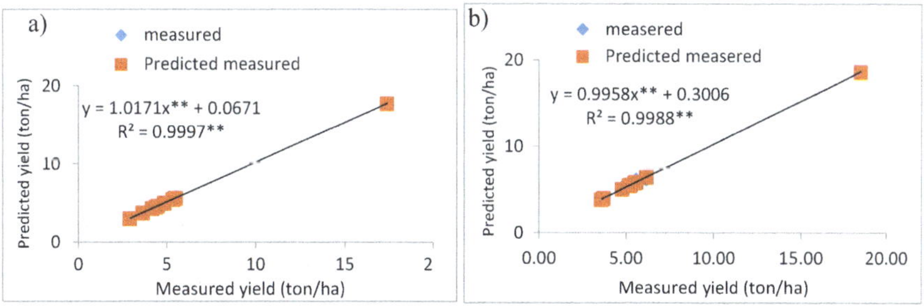

Fig. 4.2 Line fit between the measured yield of the studied crops and its corresponded simulated values by SALTMED model in the first growing season (**a**) and in the second growing season (**b**)

Nitrogen Uptake and Leaching as Affected by the Inclusion of Legume Crops in the Studied Crop Sequences

SALTMED model simulates the plant nitrogen uptake and nitrogen leaching. Although there were no field measurements for them in the published paper by Ouda et al. (2020), we presented theses simulated results to interpret the increase in the yield of studied crops under legume crops inclusion in the crop sequence.

Nitrogen Uptake by the Crops Cultivated in CS1 and Nitrogen Leaching

As it was stated earlier, CS1 was composed of maize, fallow then wheat (a farmer practice). This crop sequence was characterized by applying a larger amount of irrigation water. The amount of nitrogen fertilizer presented in Table 4.1 was applied by broadcasting over the soil surface. It is expected that high leaching of the applied fertilizer to the groundwater will occur.

Figure 4.3 presents the nitrogen uptake by maize and wheat in CS1 in the first and the second growth seasons. The figure showed that the accumulated nitrogen uptake of wheat was higher than that of maize in both growing seasons, which could be attributed to larger applied irrigation water to maize, compared to the amount applied to wheat. Furthermore, the accumulated nitrogen uptake was slightly higher in the second growing season in maize, compared to the first growing season, as a result of lower applied water in the second growing season. Whereas, for wheat, it was almost the same in both growing seasons.

Figure 4.4 presented the nitrogen leaching in the end of each growing season. The figure showed that the total nitrogen leaching was higher for maize than that of wheat. It was lower in the second growing season than its value in the first growing season for maize due to higher applied irrigation water in the first growing season than that of the second growing season. The total nitrogen leaching was higher for wheat in the second growing season due to higher applied irrigation water in the second growing season than that of the first growing season. Zafar and Muhammad (2007) reported that about 50% of applied nitrogen fertilizer remains unavailable to a crop due to temporary immobilization in soil organic matter or due to losses by leaching, erosion nitrification or volatilization.

The above result implied that, in this type of soil, namely the sandy soil, it will face a decline in its nutrients status as a result of improper management practices as in CS1, and that will constitute one of the basic limitations to food production.

Furthermore, SALTMED model is capable to simulate nitrogen uptake and nitrogen leaching in the first and the second growing seasons together and present each of them in one graph. Figures 4.5 and 4.6 showed the nitrogen uptake and nitrogen leaching during the two growing seasons. Both graphs showed similar trends as the one found in each individual growing season.

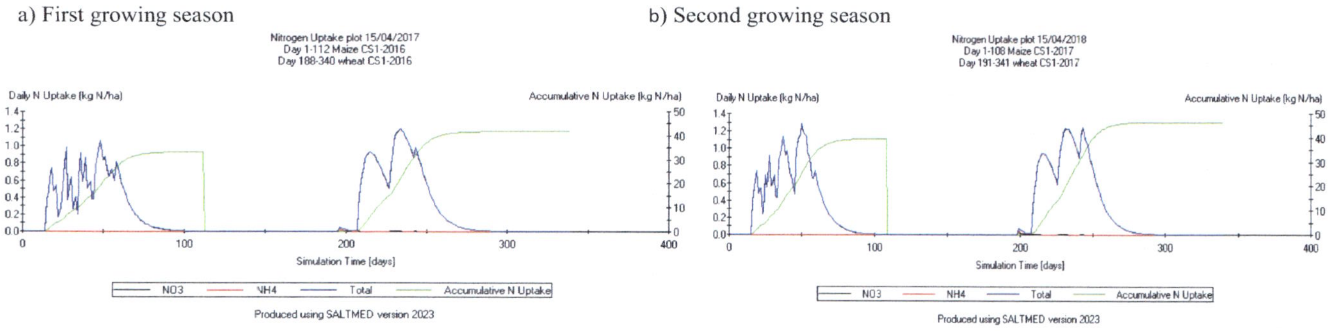

Fig. 4.3 Nitrogen uptake by the cultivated crops in CS1 in the first growing season (**a**) and the second growing season (**b**)

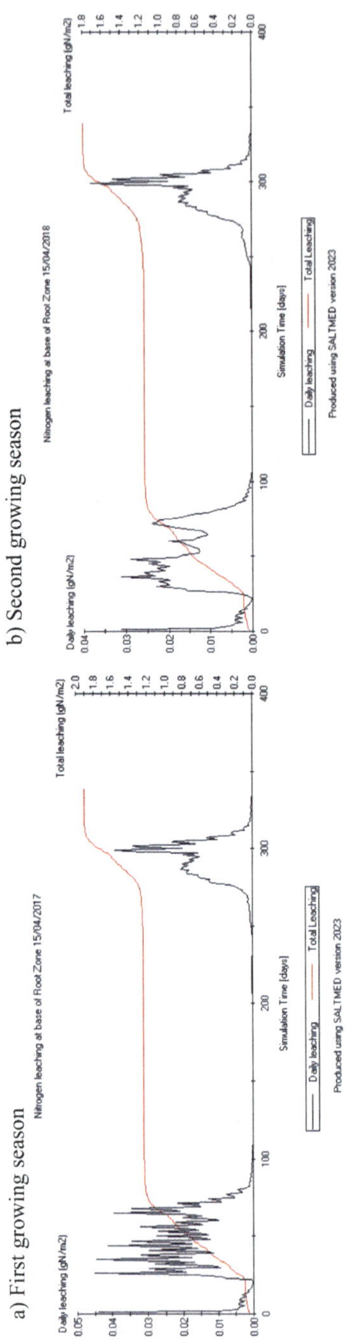

Fig. 4.4 Nitrogen leaching at the harvest of CS1 in the first growing season **(a)** and the second growing season **(b)**

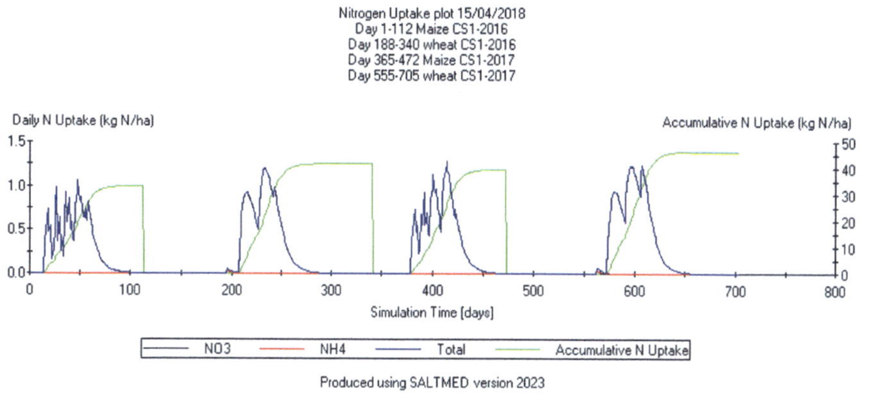

Fig. 4.5 Nitrogen uptake by the cultivated crops in CS1 in both growing seasons

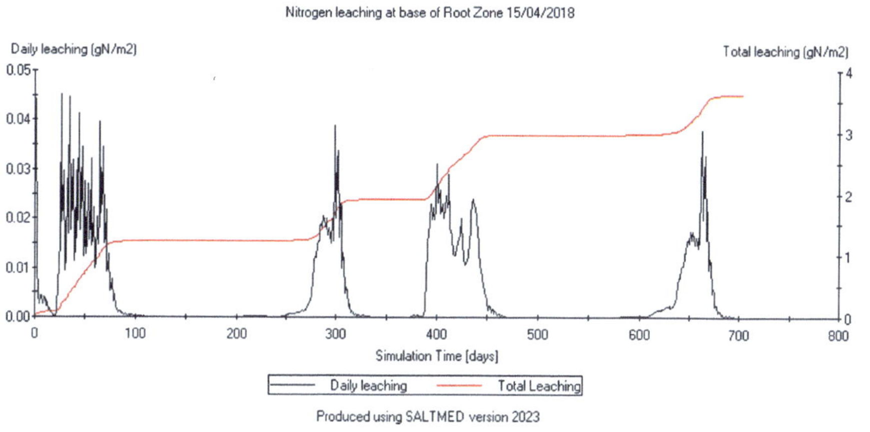

Fig. 4.6 Nitrogen leaching at the base of the soil after the harvest of CS1 in both growing seasons

Nitrogen Uptake by the Crops Cultivated in CS2 and Nitrogen Leaching

The CS2 is composed of maize, fahl clover then wheat. In CS2, the applied irrigation water was lower than its value in CS1 by an average of 23 and 22% in maize and wheat, respectively. Because the fertilizer was applied in CS2 via fertigation pump, the simulation of SALTMED model of nitrogen uptake and nitrogen leaching in CS2 was different than what was found in CS1 (Figs. 4.7 and 4.8). Figure 4.7 revealed a noticeable increase in the accumulated nitrogen uptake for both maize and wheat, due to lower applied irrigation water (as the required irrigation water). In addition, fahl clover preceded wheat, which could increase the residual nitrogen in the soil after the its harvest and that could increase nitrogen uptake by wheat. Moreover, higher values of accumulated nitrogen uptake in maize and wheat were found in the second growing season, which imply that maize and wheat in CS2 were benefited from the residual effect of fahl clover from the first growing season. The SALTMED model illustrated that in Fig. 4.8a, b, where a high values of soil nitrogen content were found after the harvest of fahl clover within the soil plot between 0.40 and 0.50 m depth in the first growing season and between 0.20 and 0.30 m depth in the second growing season.

Figure 4.9 show a slightly lower nitrogen leaching during the second growing season of maize, and an obvious reduction in nitrogen leaching during the growing season of fahl clover and wheat (Fig. 4.9a), compared to its values in first growing season (Fig. 4.9b).

Osborne et al. (2010) indicated that inclusion of a legume crop in a cropping system positively influence specific microorganism populations in the rhizosphere.

Furthermore, the simulation results also showed that nitrogen uptake during the first and second growing seasons was obviously higher than its values in each individual growing season (Fig. 4.10), which implied that the simulation of multi-year cultivation produced results of slightly higher quality, compared to the simulation of single-year cultivation. This can be attributed to the ability of the successive cereal crops to better take advantage of soil residual effect of nitrogen after the harvest of fahl clover (Fig. 4.8a, b).

Figure 4.11 supported the above findings, where nitrogen leaching was lower in the multi-year cultivation, than that found in the simuation of each single year cultivation, which implied the better ability of the cultivated crops to efficintly absorb soil nitrogen.

Nitrogen Uptake by the Crops Cultivated in CS3 and Nitrogen Leaching

In CS3, two legume crops are included, namely cowpea intercropped with maize and fahl clover. The value of applied irrigation water to CS3 was as the same as the value applied to CS2 and fertigation was also used. However, the SALTMED simulation of nitrogen uptake, and nitrogen leaching, as well as soil nitrogen content after fahl clover harvest in CS3 in both growing seasons was different than that found in CS2 (Figs. 4.12, 4.13 and 4.14). Figure 4.12a revealed the accumulated nitrogen uptake of the crops in the first growing season was similar to the one found in the first growing season of CS2 (Fig. 4.13a versus Fig. 4.8a). However, in the second growing season, the nitrogen uptake by the crops was higher than its values in CS2 (Fig. 4.13b versus Fig. 4.8b). The simulation results of SALTMED model also showed that, in Fig. 4.13a, b, higher values of soil nitrogen content were found than the one existed in CS2 after the harvest of fahl clover in the soil depth between 0.40 and 0.50 m in the first growing season and between 0.20 and 0.30 m in the second growing season.

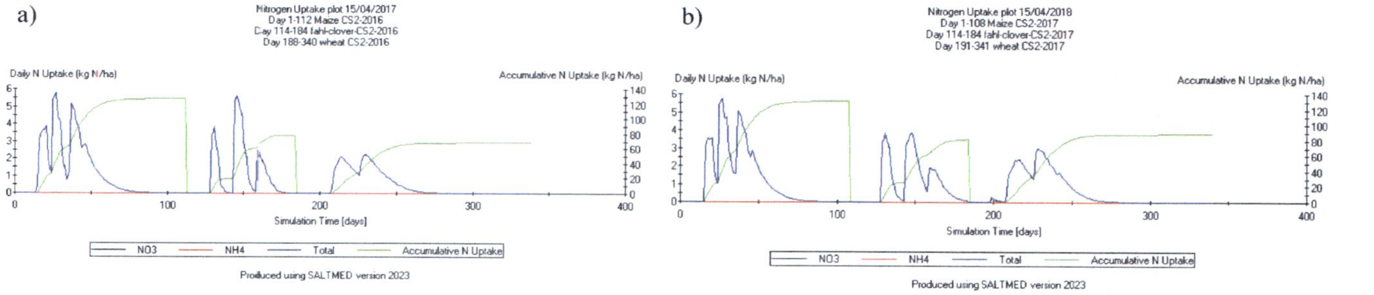

Fig. 4.7 Nitrogen uptake by the cultivated crops in CS2 in the first growing season (**a**) and the second growing season (**b**)

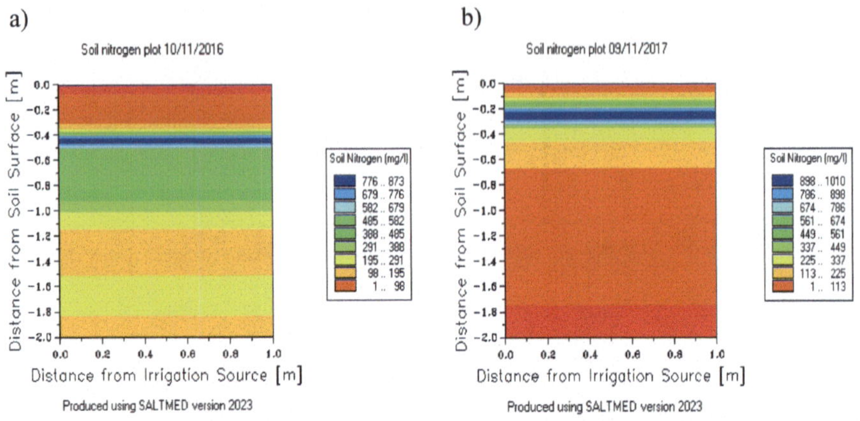

Fig. 4.8 Soil nitrogen content after cultivation of fahl clover in CS2 in the first growing season (**a**) and the second growing season (**b**)

Lower nitrogen leaching amounts were found in the second growing season (Fig. 4.14a), compared to the first growing season of CS3 (Fig. 4.14b). These results implied that the inclusion of two legume crops in the crop sequence, namely cowpea and fahl clover increased soil nitrogen, which increased its uptake by the cultivated crops and reduced the nitrogen amount that lost by leaching. Inclusion of legume crops in the cropping system can reduce the groundwater pollution by nitrogen fertilizer through make it accessible to be absorbed by the cereal crops.

Cowpea intercropping with maize system has several advantages. Kariaga (2004) concluded that cowpea intercropping systems reduces runoff through maintaining ground cover, which reduce soil erosion, as well as water evaporation and improves conservation of soil moisture (Ghanbari et al., 2010). Zohry (2005) reported an increase in maize yield by 10%, when cowpea was intercropped with maize, as a result of reduction in the associated weeds competing with maize plants. Ghanbari et al. (2010) reported an increase in the absorbed photosynthetically active radiation by cowpea in its intercropping system with maize, which positively influenced maize plants. In addition, Hamd-Allah et al. (2014) stated that reduction in biological predators that attack maize plants was observed when cowpea was intercropped with maize.

Furthermore, the simulation results also showed that nitrogen uptake during the first and second growing season of CS3 was obviously higher than its values in each individual growing season, which put an emphasis on the simulaton of multi-year cultivation, which gave more accurte results, compared to the simulation of single growing season (Fig. 4.15).

Figure 4.16 confirmed the above findings, where nitrogen leaching was lower in the multi-year cultivation, than that of each single year, which implied that the cultivated crops were more able to efficintly absorb soil nitrogen.

The role of legumes in achieving the sustainable use of soil and water resources has been extensively studied. The main benefits of legumes is increasing soil fertility

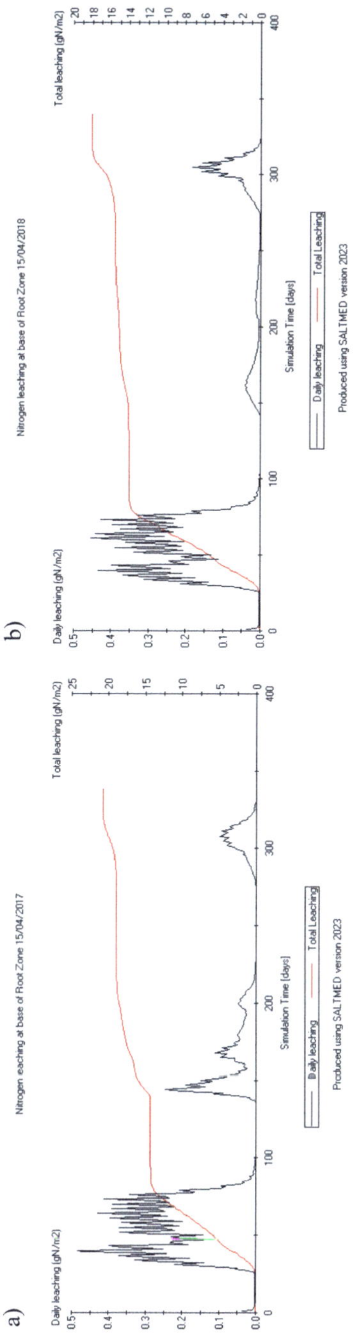

Fig. 4.9 Nitrogen leaching after the harvest of CS2 in the first growing season (**a**) and the second growing season (**b**)

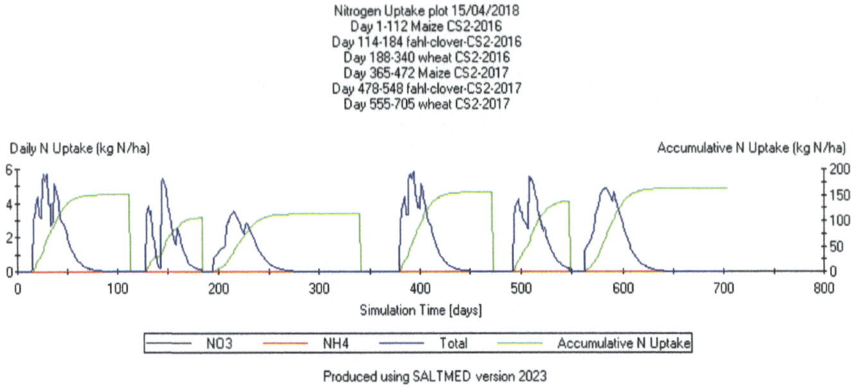

Fig. 4.10 Nitrogen uptake by the cultivated crops in CS2 in both growing seasons

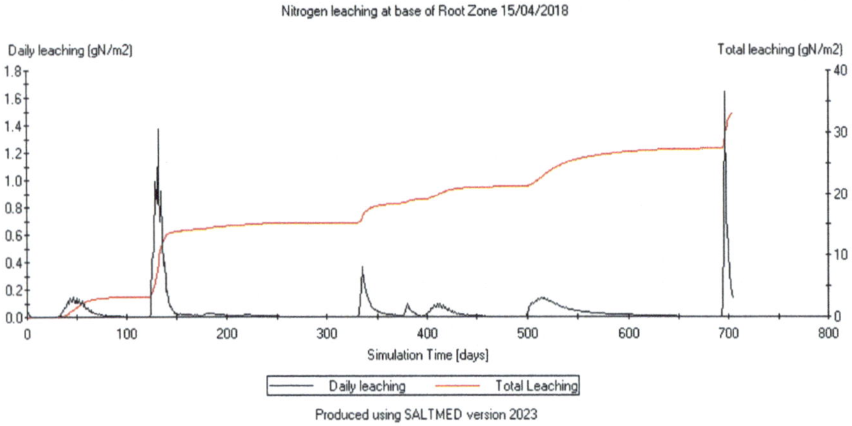

Fig. 4.11 Nitrogen leaching after the harvest of CS2 in both growing seasons

through the action of microorganisms, which improve the soil biological, chemical, and physical properties (Vasconcelos et al., 2020). Nuruzzaman et al. (2005) stated that a legume crop helps in the introduction of atmospheric nitrogen into agricultural systems through symbiotic nitrogen fixation, which has the potential to enhances soil nitrogen for subsequent cereal crop (Ruisi et al., 2017). Furthermore, through their nitrogen fixing capabilities, legumes have the ability to restore soil fertility and break the cycles of diseases and other pests attacking crops (Kebede, 2020). McCallum et al. (2004) indicated that inclusion of legumes in a cropping sequence can improve soil porosity, and structure for the benefit of following crops.

Additionally, intercropping with legumes allows the production of high yields with lower inputs of nitrogen fertilizer (Li et al., 2020). Symbiotic nitrogen fixation performed in the nodules of legume roots is a key agro-ecosystem service provided by legumes, which increases soil nitrogen availability for intercropped

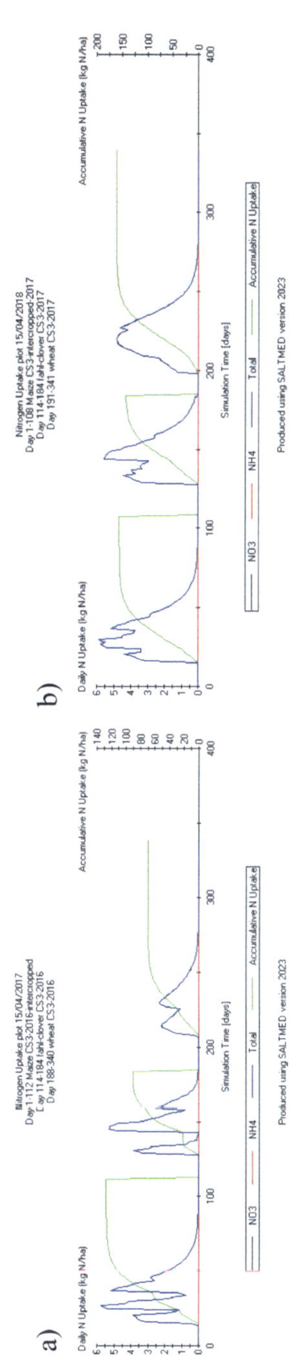

Fig. 4.12 Nitrogen uptake by the cultivated crops in CS3 in the first growing season (**a**) and the second growing season (**b**)

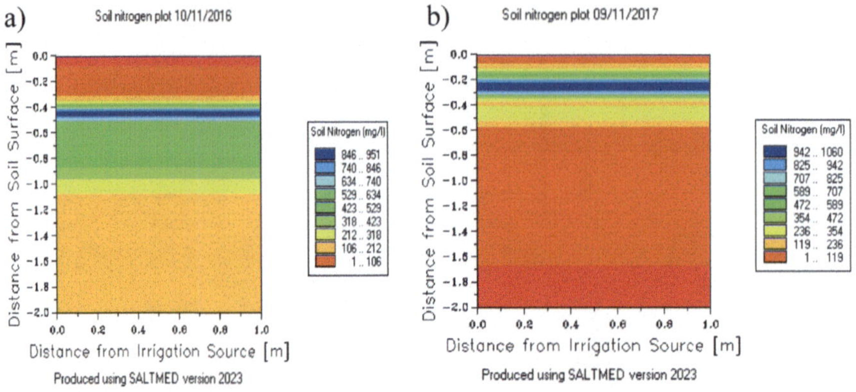

Fig. 4.13 Soil nitrogen content after cultivation of fahl clover in CS3 in the first growing season (**a**) and the second growing season (**b**)

crops (Ivanov et al., 2012). Dwivedi et al. (2015) reported that the deep roots of legume crops penetrate far into the soil and use nutrients and moisture from deeper soil layers. Consequently, a higher total nitrogen uptake per unit of land in intercrops is obtained, compared to the values obtained under sole cropping (Lithourgidis & Dordas, 2010). According to Mayer et al. (2003), nitrogen derived from legume rhizodeposits contribute to an increase of 35–44% in residual nitrogen content in the soil and constituted 79–85% of the below-ground nitrogen of plants at the maturity stage.

Legumes-based cropping systems improve several aspects of soil fertility. Stagnari et al. (2017) reported that soil organic carbon and humus content, nitrogen and phosphorous availability and uptake are improved after legumes cultivation. Furthermore, it increases organic carbon and nitrogen, as well as releasing hydrogen gas as a by-product of biological nitrogen fixation, which promotes bacterial legume nodules' development in the rhizosphere (Li et al., 2020). In addition, Ferguson et al. (2013) indicated that legumes have the ability to remove calcium and magnesium in the soil more than cereals and replace it with hydrogen, which results in removing OH^- ions and increases H^+ thus lowering the soil pH.

Intercropping a C3 crop (cowpea) with a C4 crop (maize) can attain several advantages. Collins et al. (2017) reported that, in an intercropping system, sometimes the total plant population exceeds the population of the base crop in the pure stand, which facilitate more plants exploit the available resources. Thus, it allows more biomass to be produced due to more assimilated production by the intercrops together. Additionally, intercropping a C3 crop with a C4 crop will have different type of CO_2 acquisition and fixation, where C4 have more efficient use of solar energy, cause an increase in its biomass with higher light intensity (Li et al., 2012). Furthermore, the increase of both biomass and CO_2 fixation with higher CO_2 concentration in C4 are faster than that in C3, which reflect more efficient use of CO_2 in C4 plants (Wang et al., 2012). Thus, in this way, more quantity of greenhouse gas (CO_2) is

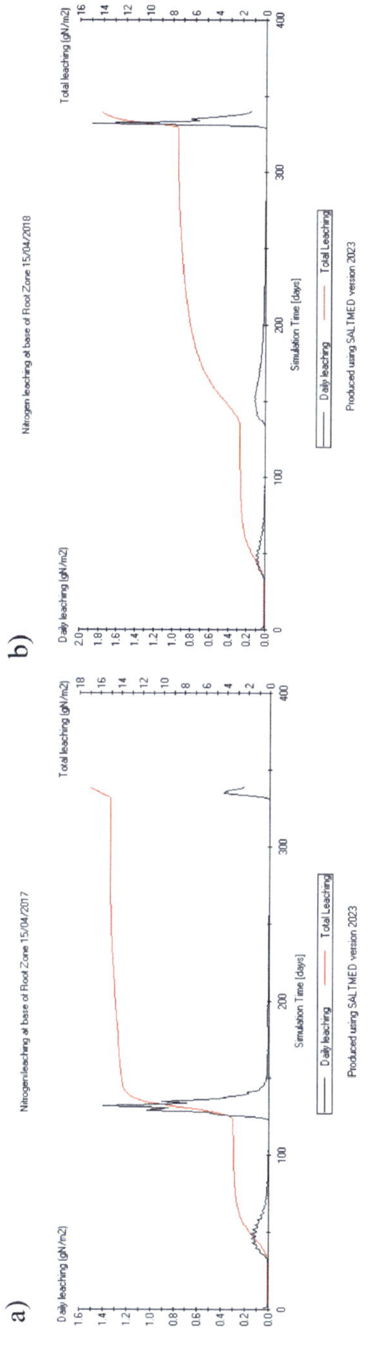

Fig. 4.14 Nitrogen leaching after the harvest of CS3 in the first growing season (**a**) and the second growing season (**b**)

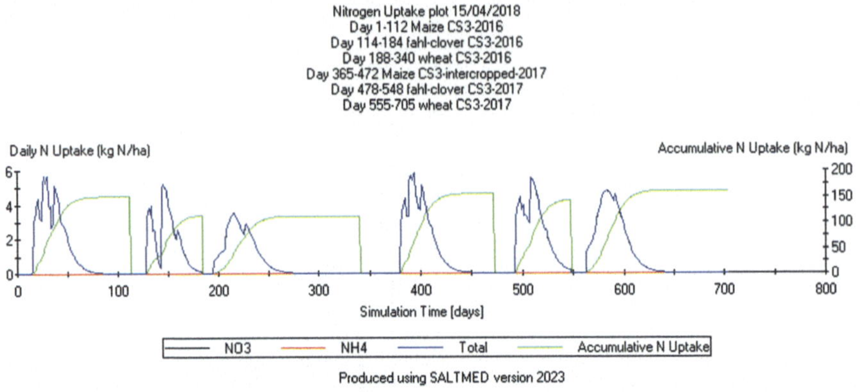

Fig. 4.15 Nitrogen uptake by the cultivated crops in CS3 in both growing seasons

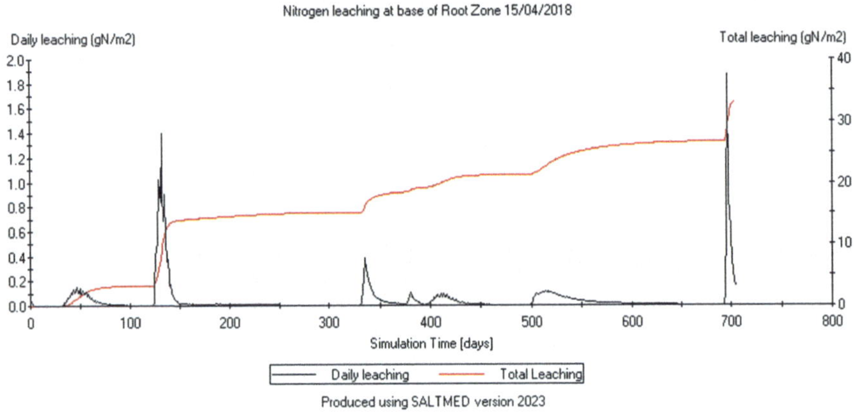

Fig. 4.16 Nitrogen leaching after the harvest of CS3 in both growing seasons

used in photosynthesis and biomass production under intercropping of C3 and C4 plants. In addition, Wu et al. (2012) stated that the taller species in an intercropping system, such as maize, could change the light and heat environment of a shorter shade-tolerant species, such as cowpea.

Legumes Increase Resiliency of Maize and Wheat to Climate Change Under Different Cropping Systems

There has been a massive body of evidences showed a continuous change in the global climatic conditions (Ali & Mujeeb-Kazi, 2021). Climate change is either directly or indirectly attributed to human activity that modifies the composition of the global atmosphere, in addition to causing the observed climate variability over comparable time periods (Karmakar et al., 2016). IPCC (2007) reported that "the global atmospheric concentration of carbon dioxide, methane and nitrous oxide has increased from pre-industrial era to 2005". Global climate models predicted 1–6 °C increase in the mean ambient temperature in the end of twenty-first century (De Costa, 2011). Global climate change not only will include increase in the atmospheric carbon dioxide concentration, but also will cause warming temperature, changing surface solar radiation, and variable precipitation, (Knox et al., 2016). In order to limit the global warning to 2 °C in 2100 to be above pre-industrial levels, annual emissions of carbon dioxide equivalents per year from the agricultural sector must be reduced by 1 giga ton by 2030 (Wollenberg et al., 2016).

The UNEP (2019) reported that climate change will increase crop water demand and decreasing crop productivity in many regions of both irrigated and rain fed agriculture across the globe. Global climate change will include increase in the atmospheric carbon dioxide concentration, warming temperature, changing surface solar radiation, and variable precipitation, which is expected to have a significant effect on crop development and production (Knox et al., 2016). Moreover, Teixeira et al. (2013) reported that in addition to that, drought, salinity, waterlogging, and mineral toxicity stresses will be also existing, which will also reduce crops production. Paymard et al. (2018) reported that not only global climate change will adversely impact crop production, but also imposes a wide range of constraints on agricultural systems, especially in water-limited environments. Furthermore, acceleration of crop development, and alteration of the phenological time duration are an important bio-indicator of climate change resulted from temperature warming (Eyshi et al., 2018).

FAO (2012) predicted growth in the world population beyond 7.5 billion inhabitants by 2030, and that will have a significant increase in water usage in irrigation by 30% and food demand will increase by 50%. In Egypt, Alkitkat (2017) projected an increase in the Egyptian population in 2030 by 32%, compared to the population value recorded in 2014. Accordingly, the demand for food is expected to increase, where Ouda and Zohry (2017) projected an increase in the demand for maize and wheat by 40 and 53%, respectively in 2030, compared to its values in 2014.

Because soils have a major role in supplying macro and micro nutrients to all kinds of crops grown in it, they are important for food security. However, climate is one of the most important factors affecting the formation of soil, thus climate change could result in soil deterioration and consequently it will threaten food security through its effects on soil properties and processes (Karmakar et al., 2016). Increased temperature is likely to have a negative effect on carbon allocation to the soil, leading to reductions in soil organic carbon (Wan et al., 2011). Increased carbon dioxide release from soils to the atmosphere will lead to more increases in the temperature (Cole et al., 2018). A decline in soil organic matter levels lead to a decrease in soil aggregate stability, infiltration rates and increase in susceptibility to compaction, run-off and susceptibility to erosion (Gorissen et al., 2004).

Avoiding soil degradation under climate change can be achieved through implementing climate-smart practices on farm level that maintain ground cover to minimize erosion and nutrients runoff (Karmakar et al., 2016). Preventing soil deterioration can be also done by the inclusion of a legume crops in the cultivated crop sequence. The sequence of crops planted in a piece of lands play an important role in increasing the productivity of the unit lands and improves soil fertility (Ouda et al., 2020). FAO (2003) stated that there is a difference between crop rotation and crop sequences. In crop rotation, different crops are grown in a fixed order year after year. But in crop sequence, the order is not necessarily repeated, and it has the flexibility in choosing the cultivated crops. Furthermore, intercropping with legumes can be an excellent practice for controlling soil degradation and sustaining crop production (Dwivedi et al., 2015). Intercropping legumes with cereals results in nitrogen fixation in the soil and consequently increase soil organic content (Hauggaard-Nielsen et al., 2009).

Climate change is expected to have large effects on global wheat production. Previous studies have revealed a decline in wheat production in the future under climate change conditions. Parry et al. (2011) reported a reduction of rain fed and irrigated wheat yield by 10–40% and 5–20%, respectively under climate change conditions. Furthermore, Liu and Yang (2010) stated that 2–5 °C increase in temperature in China could have more negative effects on wheat production than that under the rise in CO_2 concentration only. Liu et al. (2016) reported that for every 1 °C increase in temperature, global wheat yields are predicted to decline by 4–6%. Whereas, Valizadeh et al. (2014) reported a rise in temperature under climate change condition will negatively impact the grain filling growth stage causing reduction in the growing season, and yield losses in comparison to the value obtained under current weather conditions. Asseng et al. (2014) indicated that the rise in temperature under climate change conditions will decrease global wheat production by 6% for each 1 °C of further temperature increase. Whereas, Eyshi and Bannayan (2012) reported a yield reduction in wheat by 50% in the 2040–2069 periods. In another simulation study in Mexico, Hernandez-Ochoa et al. (2018) projected that the national wheat production of Mexico will decline by 7% under RCP4.5 and 8% under RCP8.5 future climate scenarios.

Likewise, the effect of climate change on maize global production has been extensively studied. Msowoya et al. (2016) stated that rain fed maize production in Malawi, may decrease up to 14% by mid-century due to climate change, rising to as much as 33% loss by the end of the century. Zhao et al. (2017) projected a reduction in maize yield by 7% for every degree increase in global mean temperature. Two studies have been done in South Africa to project the change in maize production under climate change. Kogo et al. (2019) projected a decline in maize production by as much as 8–38% under RCP4.5 and RCP8.5 scenarios by the end of the twenty-first century. Whereas, Choruma et al. (2022) projected a decrease in maize production by 24% using climate scenario RCP8.5 resulted from MIROC model in 2070–2099 period.

In this section, we used the SALTMED simulation model and CMIP6 mean projection of two future scenarios, namely SSP1-2.6 and SSP5-8.5 in two-time intervals, namely 2060–2079 and 2080–2099 to evaluate the role of legumes in enhancing the resiliency of maize and wheat to climate change when grown under different cropping systems.

Assessment of the Effect of Climate Change on the Productivity of Maize and Wheat Grown in Different Crop Sequences

Future Climate Scenarios

The data of CMIP6 mean projection of two future climate scenarios, namely SSP1-2.6 and SSP5-8.5 were collected from The World Bank Group, Climate Change Knowledge Portal (CCKP) in the following website: https://climateknowledgeportal.worldbank.org/. The CCKP serves as a leading hub of global climate data, information, and tools to support decision making and improve the understanding of future climates and projected risks. The CCKP was designed to provide open access to climate data and related information in order to help users better understand future climate scenarios and characterize potential risks and impacts with data aggregated at national, sub-national, and watershed levels (World Bank Group, 2021).

Climate projection data in that website is a modeled data, derived from the Coupled Model Inter-comparison Projects. CMIP data is the foundational data used in the IPCC Assessment Reports; CMIP6 supports the IPPC's Sixth Assessment Report, which it initial release was in August 2021. Projection data in that website is presented at a spatial resolution of $1^{\circ} \times 1^{\circ}$ (100 km × 100 km). In CMIP6, future climate scenarios are presented through five SSPs: SSP1-1.9, SSP1-2.6, SSP2-4.5, SSP3-7.0, and SSP5-8.5, which present different societal development pathways. The total radiative forcing level by 2100 (the cumulative measure of GHG emissions from all sources) is presented at the end of each pathway (i.e., − 1.9, − 2.6, − 4.5, 7.0, 8.5) (World Bank Group, 2021).

We selected two future climate scenarios to be used in this assessment, namely SSP1-2.6 and SSP5-8.5. The description of the two selected future scenarios is as followed:

SSP1-2.6: supports increasing sustainability with global emissions cut severely, but reach net-zero after 2050.

SSP5-8.5: presents a future based on an intensified exploitation of fossil fuel resources, where global markets are increasingly integrated leading to innovations and technological progress (World Bank Group, 2021).

We downloaded the projected monthly data composed of maximum and minimum temperatures as averages of Egypt and sub-region averaged over two periods, namely 2060–2079 and 2080–2099 for SSP1-2.6 and SSP5-8.5 future climate scenarios developed from a multi-model ensemble. From this data, we selected the monthly data of Minia governorate, where the studied experiment was carried out.

Calculation of Missing Climate Variables

The downloaded data of the two studied future climate scenarios contains only monthly data of maximum and minimum temperature. Whereas, the SALTMED model uses daily data, therefore the monthly data was converted to daily data using MS-Excel function "Random number generation". The model also requires to input relative humidity, sunshine hours and wind speed to calculate evapotranspiration and dry matter accumulation.

Thus, relative humidity was calculated, using the equation presented by Allen et al. (1998). The relative humidity (RH) expresses the degree of saturation of the air as a ratio of the actual vapor pressure (e_a) to the saturation vapor pressure (e_o) as followed (Allen et al., 1998):

$$e_a = 0.611 * \exp((17.27 * T_{min})/(T_{min} + 237.3)) \tag{4.5}$$

$$e_o = 0.618 * \exp((17.27 * T_{max})/(T_{max} + 237.3)) \tag{4.6}$$

$$\mathrm{RH} = 100 * (e_a/e_o) \tag{4.7}$$

where: T_{min} and T_{max} are minimum and maximum temperatures (°C), respectively.

We also calculated sunshine hours (SSH) using the equation presented by Abd El-Wahed and Snyder (2015) as followed:

$$\mathrm{SSH} = 4.352 + 0.232(T_{mean}) \tag{4.8}$$

where: T_{mean} is mean temperature (°C)

With respect to wind speed, Allen et al. (1998) stated that "the variation in wind speed average over monthly periods is relatively small and fluctuates around average values and the average wind speed estimates may be selected from information available for the regional climate". Therefore, the wind speed data under current climate was used to substitute for its unavailability in the future scenarios.

Calculation of Growing Degree Days

To determine the effect of climate change on the season length of the studied crops in the future, we used the function in SALTMED, namely "Degree days, fixed sowing date" to determine it, which involves the input of stage-specific growing degree days (GDD). This function allows the model to determine the beginning and the end of each growth stage. GDD for each growth stage of the studied crops was calculated using the following equation (Yang et al., 1995):

$$\mathrm{GDD} = T_{mean}/T_{base} \tag{4.9}$$

where: T_{base} is base temperature, which is the threshold under which plant growth ceased.

Projection of the Yield of the Studied Crops Under Future Climate Change

Future Climate Scenario SSP1-2.6 in 2060–2079

Table 4.5 presented the projection of the yield of the studied crops in the three crop sequences in both growing seasons using SSP1-2.6 in 2060–2079. The results in the table showed that the highest reduction in the yield of maize and wheat will occur under implementing CS1, which characterized by applying large amount of irrigation water and fertilizer broadcasted above the soil. The losses in the yield of these two crops will be even higher in the second growing season, compare to the first growing season as a result of soil deterioration by the continuous cultivation of these two cereal crops. In CS2, maize yield will be slightly improved as a result of implementing irrigation scheduling and fertigation in both growing seasons. Cultivation of fahl clover after maize harvest and before wheat cultivation will reduce wheat yield losses under the studied future scenario from 12% under CS1 to 9% under CS2 in the first growing season and from 15% under CS1 to 7% under CS2. Inclusion of two legume crops in CS3, namely cowpea intercropped with maize and fahl clover before, wheat cultivation will improve the yield of maize and wheat and will reduce its yield losses under the studied future climate scenario. Maize yield will be reduced by 9 and 7% in the first and second growing season, respectively. Whereas, wheat yield losses will be reduced to 6 and 4% in the first and second growing season, respectively.

The linear regression analysis showed high values of R^2 namely 0.9994 and 0.9984 in the first and second growing season, respectively. The goodness of fit analysis between measured and projected yield also showed high values of *d-stat*, namely 0.9916 and 0.9941 in the first and second season, respectively. Furthermore, RMSE values between measured and projected yield were 0.8057 and 0.7057 ton/ha in the first and second growing season, respectively, which implied high differences between the measured and projected yield of the studied crops under the SSP1-2.6 in 2060–2079 (Table 4.5).

A significant positive relationship ($P < 0.001$) was found between the measured yield of the studied crops and its corresponded projected values by the SALTMED model in both growing season (Fig. 4.17).

Table 4.5 Measured versus projected yield (ton/ha) of the studied crops in each crop sequence in the two growing seasons, and percentage of yield reduction (PR%) under SSP1-2.6 future scenario in 2060–2079

	First growing season			Second growing season		
	Measured	Projected	PR (%)	Measured	Projected	PR (%)
CS1						
Maize	5.55	4.95	11	5.52	4.86	12
Wheat	3.81	3.33	13	3.79	3.24	15
CS2						
Maize	5.68	5.09	10	5.97	5.35	10
Fahl clover	4.41	4.10	7	5.23	4.85	7
Wheat	4.37	3.98	9	5.54	5.16	7
CS3						
Cowpea/maize	17.76	16.16	9	18.69	17.30	7
Fahl clover	4.88	4.56	7	5.75	5.45	5
Wheat	4.72	4.46	6	6.16	5.93	4
R^2	0.9994			0.9984		
d-stat	0.9916			0.9941		
RMSE (ton/ha)	0.8057			0.7057		

CS1 = maize, fallow, then wheat, CS2 = maize, improved fallow (fahl clover) then wheat, CS3 = cowpea intercropped with maize, improved fallow (fahl clover) then wheat

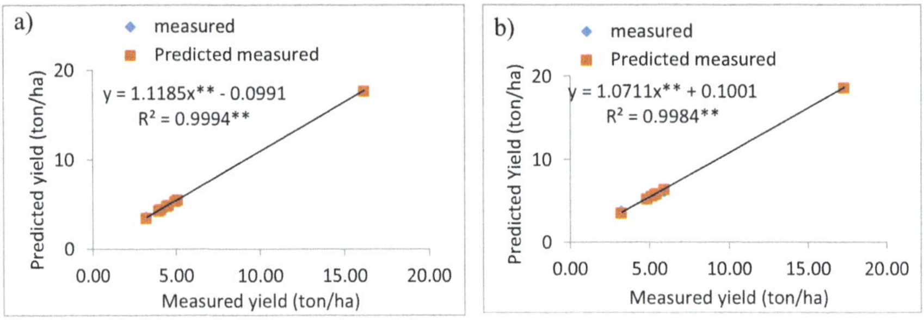

Fig. 4.17 Line fit between the measured yield of the studied crops and its corresponded projected values by SALTMED model in the first growing season (**a**) and in the second growing season (**b**) under SSP1-2.6 in 2060–2079

Future Climate Scenario SSP1-2.6 in 2080–2099

Because SSP1-2.6 describes a future with increasing sustainability and highly cut in the global emissions reaching net-zero after 2050, the reductions in the yield of the cultivated crops in the three crop sequences in 2080–2099 (Table 4.6) were close to the values obtained in 2060–2079 (Table 4.5). Additionally, the highest reduction in the yield of maize and wheat will occur under implementing CS1, where it will be higher in the second growing season, compare to the first season as a result of soil deterioration under this practice. Implementing irrigation scheduling with fertigation and inclusion of fahl clover in CS2 will reduce maize yield loss in the second growing season and wheat yield loss in both growing seasons. Likewise, implementing irrigation scheduling with fertigation and inclusion of two legume crops in CS3, will improve the yield of maize and wheat and reduce its losses under the studied climate scenario. Thus, the loss in maize yield will decrease from 10% under CS1 to 8% under CS3 averaged over the two growing seasons, whereas the loss in wheat yield will decrease from 15% under CS1 to 5% under CS3 averaged over the two growing seasons (Table 4.6).

The analysis also showed high values of R^2, namely 0.9995 and 0.9981 in the first and second growing season, respectively. The goodness of fit analysis also showed high values of *d-stat*, namely 0.9917 and 0.9941 in the first and second growing season, respectively. RMSE values between measured and projected yield were 0.8056 and 0.7069 ton/ha in the first and second growing season, respectively (Table 4.6).

Figure 4.18 showed a significant positive relationship ($P < 0.001$) between the measured yield of the studied crops and its corresponded projected values in both growing seasons.

Table 4.6 Measured versus projected yield (ton/ha) of the studied crops in each crop sequence in the two growing seasons, and percentage of yield reduction (PR%) under SSP1-2.6 future scenario in 2080–2099

	First growing season			Second growing season		
	Measured	Predicted	PR (%)	Measured	Predicted	PR (%)
CS1						
Maize	5.55	5.01	10	5.52	4.95	10
Wheat	3.81	3.26	14	3.79	3.15	17
CS2						
Maize	5.68	5.14	10	5.97	5.31	11
Fahl clover	4.41	4.09	7	5.23	4.90	6
Wheat	4.37	3.98	9	5.54	5.16	7
CS3						
Cowpea/maize	17.76	15.96	10	18.69	17.30	7
Fahl clover	4.88	4.56	7	5.75	5.45	5
Wheat	4.72	4.46	6	6.16	5.93	4
R^2	0.9995			0.9981		
d-stat	0.9917			0.9941		
RMSE (ton/ha)	0.8056			0.7069		

CS1 = maize, fallow, then wheat, CS2 = maize, improved fallow (fahl clover) then wheat, CS3 = cowpea intercropped with maize, improved fallow (fahl clover) then wheat

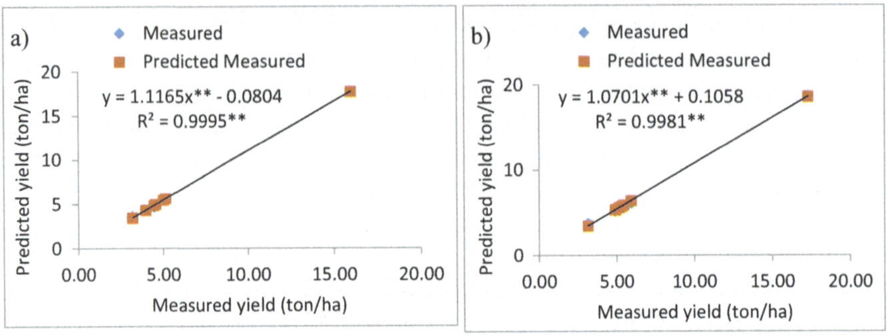

Fig. 4.18 Line fit between the measured yield of the studied crops and its corresponded projected values by SALTMED model in the first growing season (**a**) and in the second growing season (**b**) under SSP1-2.6 in 2080–2099

Future Climate Scenario SSP5-8.5 in 2060–2079

Because SSP5-8.5 represents stressful conditions, where it is based on an intensified exploitation of fossil fuel resources, it is expected that more losses in the maize and wheat yield will occur. The simulation results projected a higher reduction in the productivity of the studied crops. Table 4.7 indicated that the highest expected reduction in maize and wheat yield will occur under CS1. Whereas, the lowest expected reduction will occur in CS3. Maize yield will be reduced by 15% under CS1, but the reduction will be 10% under CS3, averaged over the two growing seasons. Similarly, wheat yield reduction will reach 15% under CS1 and it will reach 8% under CS3, averaged over the two growing seasons.

Table 4.7 also showed high values of R^2 namely 0.9991 and 0.9981 in the first and second growing season, respectively. The goodness of fit analysis showed high values of *d-stat*, namely 0.9901 and 0.9912 in the first and second growing season, respectively. RMSE values between measured and projected yield were 0.8765 and 0.8620 ton/ha in the first and second growing season, respectively.

Figure 4.19 showed a significant positive relationship ($P < 0.001$) between the measured yield of the studied crops and its corresponded projected values in both growing seasons.

Future Climate Scenario SSP5-8.5 in 2080–2099

The results in Table 4.8 projected higher reduction in the yield of the studied crops under the three crop sequences, compare to the reduction found under SSP1-2.6 in 2060–2079 and 2080–2099, as well as SSP5-8.5 in 2060–2079. Under CS1, maize yield will be reduced by 25% and wheat yield will be reduced by 23%, averaged over

Table 4.7 Measured versus projected yield (ton/ha) of the studied crops in each crop sequence in the two growing seasons, and percentage of yield reduction (PR%) under SSP5-8.5 future scenario in 2060–2079

	First growing season			Second growing season		
	Measured	Predicted	PR (%)	Measured	Predicted	PR (%)
CS1						
Maize	5.55	4.75	14	5.52	4.69	15
Wheat	3.81	3.22	15	3.79	3.20	16
CS2						
Maize	5.68	5.12	10	5.97	5.44	9
Fahl clover	4.41	4.09	7	5.23	4.88	7
Wheat	4.37	3.92	10	5.54	5.11	8
CS3						
Cowpea/maize	17.76	15.86	11	18.69	16.91	10
Fahl clover	4.88	4.56	7	5.75	5.43	6
Wheat	4.72	4.28	9	6.16	5.73	7
R^2	0.9991			0.9981		
d-stat	0.9901			0.9912		
RMSE (ton/ha)	0.8765			0.8520		

CS1 = maize, fallow, then wheat, CS2 = maize, improved fallow (fahl clover) then wheat, CS3 = cowpea intercropped with maize, improved fallow (fahl clover) then wheat

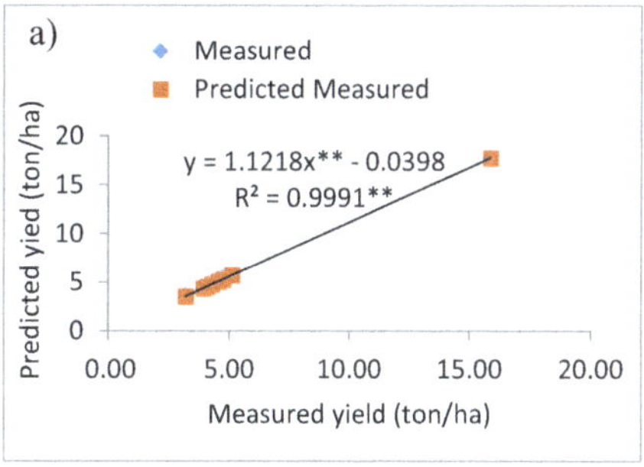

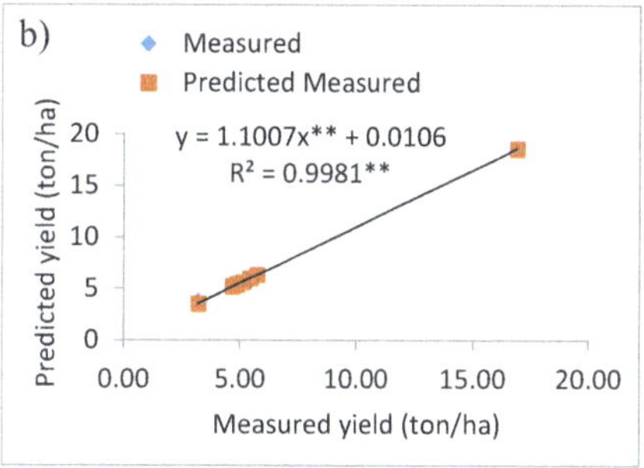

Fig. 4.19 Line fit between the measured yield of the studied crops and its corresponded projected values by SALTMED model in the first growing season (**a**) and in the second growing season (**b**) under SSP5-8.5 in 2060–2079

the two growing seasons. Whereas, under CS3, maize yield will be reduced by 21% and wheat yield will be reduced by 17%, averaged over the two growing seasons.

Table 4.8 also showed high values of R^2 namely 0.9979 and 0.9967 in the first and second growing season, respectively. Furthermore, the goodness of fit analysis showed high values of *d-stat*, namely 0.9540 and 0.9622 in the first and second growing season, respectively. RMSE values between measured and projected yield

Table 4.8 Measured versus projected yield (ton/ha) of the studied crops in each crop sequence in the two growing seasons, and percentage of yield reduction (PR%) under SSP5-8.5 future scenario in 2080–2099

	First growing season			Second growing season		
	Measured	Predicted	PR (%)	Measured	Predicted	PR (%)
CS1						
Maize	5.55	4.19	25	5.52	4.17	24
Wheat	3.81	2.99	22	3.79	2.95	22
CS2						
Maize	5.68	4.45	22	5.97	4.81	19
Fahl clover	4.41	3.43	22	5.23	4.38	16
Wheat	4.37	3.49	20	5.54	4.68	16
CS3						
Cowpea/maize	17.76	13.83	22	18.69	15.06	19
Fahl clover	4.88	4.22	14	5.75	5.03	13
Wheat	4.72	3.81	19	6.16	5.24	15
R^2	0.9979			0.9967		
d-stat	0.9540			0.9622		
RMSE (ton/ha)	1.7947			1.6851		

CS1 = maize, fallow, and wheat, CS2 = maize, improved fallow (fahl clover) then wheat, CS3 = cowpea intercropped with maize, improved fallow (fahl clover) then wheat

were 1.7947 and 1.6851 ton/ha in the first and second growing season, respectively (Table 4.8), which reflect the high differences between the measured and projected values.

The linear regression analysis showed a significant relationship ($P < 0.001$) between the measured yield values of the studied crops and its corresponded projected values in both growing seasons (Fig. 4.20).

In Egypt, maize production has been projected under future climate change by several authors using simulation model, such as CropSyst model. Ouda et al. (2009) indicated that maize grown in clay soil under surface system will experience high yield losses, namely 49 and 47% under A2 and B2, respectively. Whereas, in silty clay soil and under drip system, maize yield losses will be lower and could reach 28 and 25% under A2 and B2 climate change scenarios, respectively (Ouda et al., 2012a). Furthermore, Morsy (2015) simulated the effect of climate change on maize yield using RCP6.0 climate change scenario developed by CCSM4 model in 2030 and projected 13–15% reduction in maize yield.

Previous studies in Egypt on the effect of climate change on wheat revealed that under A2 and B2 climate change scenarios developed from HadCM3 model and CropSyst simulation model, Moneir (2012) found that for wheat planted in clay soil and under surface irrigation, yield losses will occur, namely 41–43%. Whereas, for wheat grown in silty clay soil and under sprinkler system, yield losses will be 39

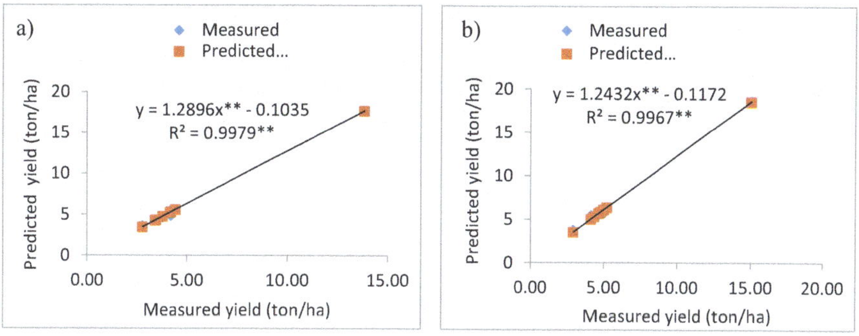

Fig. 4.20 Line fit between the measured yield of the studied crops and its corresponded projected values by SALTMED model in the first growing season (**a**) and in the second growing season (**b**) under SSP5-8.5 in 2080–2099

and 34% under A2 and B2 climate change scenarios, respectively (Noreldin et al., 2012). In salt-affected soil and under surface irrigation, wheat yield losses under farmer irrigation practice (high applied irrigation water) will be high, namely 47 and 44% under A2 and B2 climate change scenarios, respectively (Ouda et al., 2012b). Similarly, for farmer irrigation practice in sandy soil and under sprinkler irrigation, yield losses will reach 32 and 27% under A2 and B2 climate change scenarios, respectively (Ouda et al., 2010). Furthermore, Morsy (2015) simulated the effect of climate change on wheat yield using RCP6.0 climate change scenario developed by CCSM4 model in 2030 and projected 10–12% decline in wheat productivity.

It has been reported that high temperature causes decline in the yields of main food crops, which is a major concern for falling agricultural productivity (Hatfield & Prueger, 2015). The negative effect of high temperature on wheat productivity is attributed directly or indirectly to the disorder in different cellular, and physiological functions, as well as metabolic pathways associated with the grain yield (Pask et al., 2014). One of the reasons for yield losses under heat stress is the premature initiation of leaf senescence by heat stress causing chlorophyll devastation, as well as inefficient nutrients recycling causing massive yield losses (Vijayalakshmi et al., 2010).

It has been also reported that the high temperature in the Mediterranean region is an inevitable constraint for wheat in the reproductive stage, where this stage is highly sensitive to heat stress (Akter & Islam, 2017). Increased temperature by 3–4 °C above the optimum temperature during wheat grain filling reduces wheat yield by 10–50% in Asia (Hussain et al., 2018). Whereas, globally, Asseng et al. (2015) reported that for each degree increase in the temperature at the grain filling duration, a reduction in wheat yield by 6% could occur.

Similarly, it has been reported that brief or prolonged episodes of high temperature during the maize growth cycle (especially at flowering stage) cause metabolic and/or morphological alterations, where it can leads to irreversible yield reductions (Sánchez et al., 2014), due to decline in maize leaf growth at temperatures above 35 °C (Shim et al., 2017), and suppresses maize ovary fertilization and the grain filling process (Lizaso et al., 2018), which is directly associated with the final grain yield. Furthermore, Alam et al. (2017) indicated that temperature of 35 °C negatively affect maize plants and reduces grain yield by up to 31% due to decreased grain number and grain weight. Temperatures ranging from 33 to 36 °C during pre-and post-flowering stages of maize, reduce crop growth rate, and grain number and grain yield (Neiff et al., 2016).

Projection of Length of the Growing Season of the Studied Crops Under Future Climate

Future Climate Scenario SSP1-2.6 in 2060–2079

The results in Table 4.9 presented the reduction in the season length of the studied crops of the three crop sequences. In CS1, maize season length will be reduced by 10 days and wheat season length will be reduced by 27 days, averaged over the two growing seasons. In CS2, longer season length was projected for both maize and wheat, where maize season length will be reduced by 10 days, and wheat season length will be reduced by 20 days, averaged over the two growing seasons. Whereas, in CS3, the longest season length was found, where maize season length will be reduced by 11 days and wheat season length will be reduced by 17 days, averaged over the two growing seasons.

Table 4.9 Measured versus projected season length (days) of the studied crops in each crop sequence in the two growing seasons, and the difference between the measured and projected season length (days) under SSP1-2.6 future scenario in 2060–2079

	First growing season			Second growing season		
	Measured	Projected	Difference	Measured	Projected	Difference
CS1						
Maize	111	100	11	107	98	9
Wheat	152	123	29	150	125	25
CS2						
Maize	111	101	10	107	97	10
Fahl clover	70	59	11	70	62	8
Wheat	152	130	22	150	132	18
CS3						
Cowpea/maize	111	101	10	107	95	12
Fahl clover	70	56	14	70	59	11
Wheat	152	132	20	150	137	13

CS1 = maize, fallow, then wheat, CS2 = maize, improved fallow (fahl clover) then wheat, CS3 = cowpea intercropped with maize, improved fallow (fahl clover) then wheat

Table 4.10 Measured versus projected season length (days) of the studied crops in each crop sequence in the two growing seasons, and the difference between the measured and projected season length (days) under SSP1-2.6 future scenario in 2080–2099

	First growing season			Second growing season		
	Measured	Projected	Difference	Measured	Projected	Difference
CS1						
Maize	111	99	12	107	96	11
Wheat	152	122	30	150	125	25
CS2						
Maize	111	100	11	107	97	10
Fahl clover	70	61	9	70	62	8
Wheat	152	130	22	150	132	18
CS3						
Cowpea/maize	111	101	10	107	95	12
Fahl clover	70	61	9	70	62	8
Wheat	152	132	20	150	137	13

CS1 = maize, fallow, then wheat, CS2 = maize, improved fallow (fahl clover) then wheat, CS3 = cowpea intercropped with maize, improved fallow (fahl clover) then wheat

Future Climate Scenario SSP1-2.6 in 2080–2099

Table 4.10 showed slightly different results than what was found in Table 4.9. The results in Table 4.10 showed that in CS1, maize season length will be reduced by 12 days and wheat season length will be reduced by 28 days, averaged over both growing seasons. In CS2, longer season length was projected for both maize and wheat, where maize and wheat season length will be reduced by 11 and 20 days, respectively averaged over both growing seasons. Whereas, in CS3, the longest season length was found, where maize and wheat season length will be reduced by 11 and 17 days, respectively averaged over both growing seasons.

Future Climate Scenario SSP5-8.5 in 2060–2079

Shorter season length will be found under SSP5-8.5 in 2060–2079. In CS1, maize season length will be reduced by 12 days and wheat season length will be reduced by 28 days, averaged over both growing seasons. In CS2, longer season length was projected for both maize and wheat, where maize and wheat season length will be reduced by 12 and 21 days, respectively averaged over both growing seasons. The longest season length was found in CS3, where maize season length was reduced by 11 days and wheat season length will be reduced by 18 days, averaged over both growing seasons (Table 4.11).

Table 4.11 Measured versus projected season length (days) of the studied crops in each crop sequence in the two growing seasons, and the difference between the measured and projected season length (days) under SSP5-8.5 future scenario in 2060–2079

	First growing season			Second growing season		
	Measured	Projected	Difference	Measured	Projected	Difference
CS1						
Maize	111	100	11	107	95	12
Wheat	152	123	29	150	123	27
CS2						
Maize	111	100	11	107	95	12
Fahl clover	70	59	11	70	62	8
Wheat	152	129	23	150	132	18
CS3						
Cowpea/maize	111	100	11	107	97	10
Fahl clover	70	56	14	70	59	11
Wheat	152	131	21	150	135	15

CS1 = maize, fallow, then wheat, CS2 = maize, improved fallow (fahl clover) then wheat, CS3 = cowpea intercropped with maize, improved fallow (fahl clover) then wheat

Future Climate Scenario SSP5-8.5 in 2080–2099

The shortest season length will be found under SSP5-8.5 in 2080–2099. In CS1, maize and wheat season length will be reduced by 25 and 42 days, respectively averaged over the two growing seasons. In CS2, longer season length will be projected for both maize and wheat, where it will be reduced by 21 and 38 days, respectively averaged over the two growing seasons. Whereas, in CS3, the longest season length will be found, where maize and wheat season length will be reduced by 24 and 37 days, respectively, averaged over the two growing seasons (Table 4.12).

It has been reported that higher temperatures accelerate crop growth, which reduces the time for vegetative and reproductive development, which limit the achievement of yield potential (Schauberger et al., 2017). Reduced crop yield under elevating temperatures is related to reducing the length of the growth stages, especially the grain filling stage (White & Reynolds, 2003). Under rainfed conditions, higher temperature could accelerate vegetative development to reach the reproductive development, where the plants requiring the most water and that could result in low yield (Harrison et al., 2011). Ahamed et al. (2010) reported a decrease in the maturity growth stage and grain filling duration of wheat under heat stress. Hastening crop development by elevated temperatures reduces the amount of solar radiation received by the plant during each developmental stage (Porter & Semenov, 2005). This led to less interception of solar energy, which will a problem when aggregated over the entire growing period (Harrison et al., 2011). Less interception of solar energy reduces photosynthesis and reduce plant biomass, and consequently reduce yield (Scott et al., 2009).

Table 4.12 Measured versus projected season length (days) of the studied crops in each crop sequence in the two growing seasons, and the difference between the measured and projected season length (days) using SSP1-2.6 future scenario in 2080–2099

	First growing season			Second growing season		
	Measured	Projected	Difference	Measured	Projected	Difference
CS1						
Maize	111	100	11	107	98	9
Wheat	152	123	29	150	125	25
CS2						
Maize	111	101	10	107	97	10
Fahl clover	70	59	11	70	62	8
Wheat	152	130	22	150	132	18
CS3						
Cowpea/maize	111	101	10	107	95	12
Fahl clover	70	56	14	70	59	11
Wheat	152	132	20	150	137	13

CS1 = maize, fallow, then wheat, CS2 = maize, improved fallow (fahl clover) then wheat, CS3 = cowpea intercropped with maize, improved fallow (fahl clover) then wheat

Effect of Climate Change on Nitrogen Use Efficiency (NUE) of the Studied Crops

Nitrogen is a major macronutrient often limiting plant growth, in which its availability has impacts throughout crop development, and on the final yield as well (Hawkesford et al., 2013). Crop responses to nitrogen is affected by the rate and timing of nitrogen application in relation to different factors, such as plant development, cultivar and climatic conditions (Masaka, 2006). Application of split doses of nitrogen during plant's growing season are known to be effective in increasing nitrogen use efficiency, rather than one single application prior to cultivation (Ooro et al., 2011).

In our study, the value of NUE was calculated as nitrogen removed with yield divided by the applied nitrogen amounts (Dobermann, 2007), which we found as the best method for comparing the different management practices between the three crop sequences. Because the yield of some of the crops in CS2 and CS3 is biological yield (fahl clover, and cowpea intercropped with maize), we made the NUE units to be kg of yield, either grains or biological yield, per kg of nitrogen (kg Y/kg N) (Fig. 4.21).

The figure showed that the values of NUE of fahl clover were the highest due to low nitrogen fertilizer amount (Table 4.1) and high biological yield (Table 4.2) under the measured and projected yield values. Likewise, the values of NUE of cowpea intercropped with maize were also high as a result high biological yield of both crops under the measured and projected yield values. Furthermore, the highest values of NUE were found for the measured values under the experimental conditions,

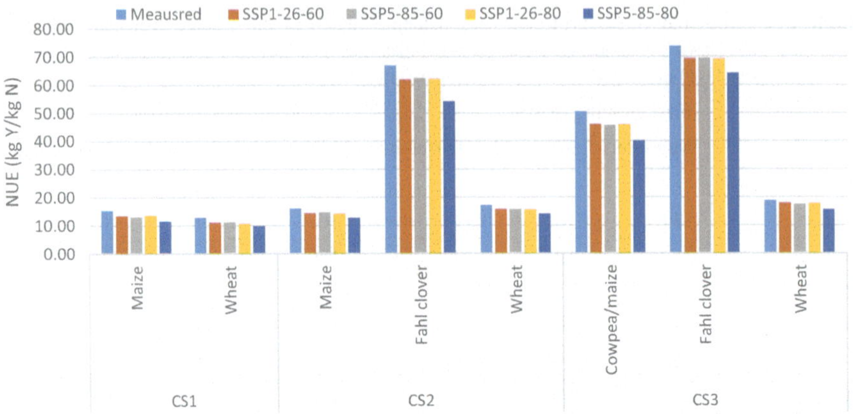

Fig. 4.21 Measured versus projected nitrogen use efficiency (NUE) resulted from the four studied future scenarios averaged over the two growing seasons

followed by the values obtained under SSP1-2.6 in 2060–2079, where yield losses of the studied crops were the lowest, compared to the measured yield values. Whereas, the lowest values of NUE were found under SSP5-8.5 in 2080–2099, where yield losses of the studied crops were the highest, compared to the measured yield values.

Effect of Climate Change on Water Productivity (WP) of the Studied Crops

Valipour (2014) defined water productivity as the ratio of yield or marketable product to water used by the crop. Figure 4.22 showed that WP values were the lowest under CS1, where irrigation was applied without scheduling under the measured and projected yield values. Furthermore, the values of WP were high for fahl clover as a result of high aboveground biomass and low applied irrigation water in both CS2 and CS3 under the measured and projected yield values. The highest values of WP were obtained for cowpea intercropping system with maize in CS3, due to high aboveground biomass and using the amount of applied water to maize to irrigate both cowpea and maize under the measured and projected yield values.

The highest values of WP were found for the measured values under the experimental conditions, followed by the values obtained under SSP1-2.6 in 2060–2079, where yield losses of the studied crops were the lowest. Whereas, the lowest values of WP were found under SSP5-8.5 in 2080–2099, where yield losses of the studied crops were the highest, compared to the measured yield values.

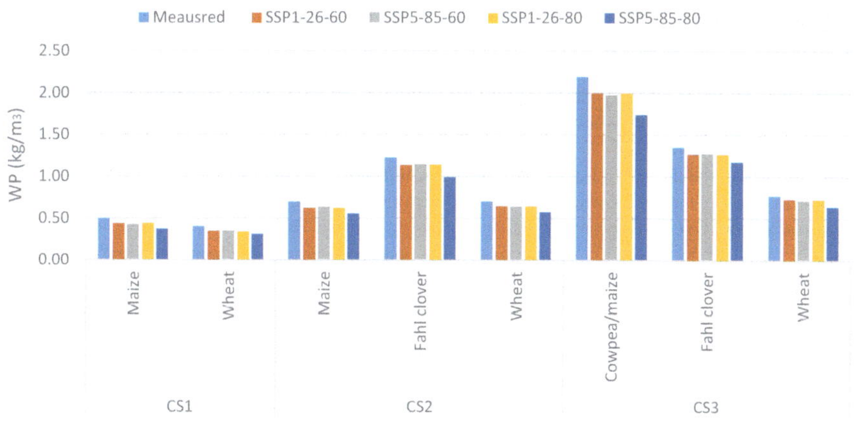

Fig. 4.22 Measured versus projected water productivity (WP) resulted from the four studied future scenarios averaged over the two growing seasons

Conclusion

In this chapter, we ascertained the advantage of the inclusion of two legume crops in maize and wheat cropping system to increase the diversification of the cropping systems, which could allow to replace agro-chemicals and fossil fuels with more ecological practices to increase food security. Using SALTMED model in the simulation of multi-year cultivation of each crop sequences allowed to examine nitrogen uptake by each crop sequence and nitrogen leaching to the soil during both growing seasons. Furthermore, the model was able to simulate the residual nitrogen content in the soil after the harvest of fahl clover, which could have been used by wheat plants in CS2 and CS3 and contributed in the high yield of wheat. Consequently, the inclusion of legumes in the cereal cropping system can reduce the need to nitrogen fertilizers, thus lower the emission of greenhouse gases, such as carbon dioxide and nitrous oxide.

Human-induced increases in atmospheric greenhouse gas concentrations have led to rising global temperatures. Thus, the increasing attention of low-input crop production and sustainable agricultural systems encourage the inclusion of legume crops in the cereal cropping systems. The simulation study implemented in this chapter proved that the inclusion of fahl clover in CS2 and cowpea and fahl clover in CS3 increase the resiliency of maize and wheat under the two future climate scenarios and reduce its yield losses, as well as improved its nitrogen use efficiency and water productivity. Therefore, inclusion of legumes crops in cereal-based cropping system can attain the sustainable use of natural resources under current weather and in the future.

References

Abd El-Wahed, M. H., & Snyder, R. (2015). Calculating sunshine hours and reference evapotranspiration in arid regions when solar radiation data are limited. *Irrigation and Drainage*. https://doi.org/10.1002/ird.1920

Ahamed, K. U., Nahar, K., & Fujita, M. (2010). Sowing date mediated heat stress affects the leaf growth and dry matter partitioning in some spring wheat (Triticum aestivum L.) cultivars. *The IIOAB Journal, 1*(3), 1–9.

Akter, N., & Islam, M. R. (2017). Heat stress effects and management in wheat. A review. *Agronomy and Sustainable Development, 37*, 1–19. https://doi.org/10.1007/s13593-017-0443-9

Alam, M. A., Seetharam, K., Zaidi, P. H., Dinesh, A., Vinayan, M. T., & Nath, U. K. (2017). Dissecting heat stress tolerance in tropical maize (*Zea mays* L.). *Field Crops Research, 204*, 110–119. https://doi.org/10.1016/j.fcr.2017.01.006

Ali, N., & Mujeeb-Kazi, A. (2021) Food production: Global challenges to mitigate climate change. In *Physiological, molecular, and genetic perspectives of wheat improvement* (pp. 1–13). Springer. https://doi.org/10.1007/978-3-030-59577-7_1

Alkitkat, H. (2017). Projections of the Egyptian population. In *Future of food gaps in Egypt: Obstacles and opportunities*. Springer Publishing House. ISBN 978-3-319-46942-3.

Allen, R. G., Pereira, L. S., Raes, D., & Smith, M. (1998). *Crop evapotranspiration: Guideline for computing crop water requirements*. FAO No. 56, pp. 109–158.

Arslan, A., Majid, G. A., Abdallah, K., Rameshwaran, P., Ragab, R., Singh, M., & Qadir, M. (2015). Evaluating the productivity potential of chickpea, lentil and faba bean under saline water irrigation systems. *Irrigation and Drainage*. https://doi.org/10.1002/ird.1912

Asseng, S., Ewert, F., Martre, P., Rotter, R. P., Lobell, D., Cammarano, D., Kimball, B., Ottman, M., Wall, G., & White, J. W. (2015). Rising temperatures reduce global wheat production. *Nature Clinical Practice Endocrinology & Metabolism, 5*, 143–147. https://doi.org/10.1038/nclimate2470

Asseng, S., Ewert, F., Martre, P., Rotter, R. P., Lobell, D. B., Cammarano, D., Kimball, B. A., Ottman, M. J., Wall, G. W., & White, J. W. (2014). Rising temperatures reduce global wheat production. *Nature Clinical Practice Endocrinology & Metabolism, 5*, 143–147. https://doi.org/10.1038/nclimate2470

Awad M. M., A. El-Mehy and H. H. Hassan. 2023. Impact of crop sequence and nitrogen fertilization levels on wheat (Triticum aestivum L.) productivity in Egypt. *Egyptian Journal of Agricultural Research, 101*(1), 83–94.

Bado, B. V., Bationo, A., & Cescas, M. P. (2006). Assessment of cowpea and groundnut contributions to soil fertility and succeeding sorghum yields in the Guinean savannah zone of Burkina Faso (West Africa). *Biology and Fertility of Soils, 43*, 171–176.

Bakheit, B. R., Abo-Elwafa, A., Abdel-Galil, M. M., & Abdelmonem, A. M. A. (2016). Impacts of recurrent selection and synthetic population on forage and seed yields of monocut Egyptian Clover (Trifolium alexandrinum L.). *Assiut Journal of Agriculture Science, 47*(1), 31–44.

Bationo, A., Hartemink, A., Lungu, O., Naimi, M., Okoth, P., Smaling, E., & Waswa, B. (2012). Knowing the African soils to improve fertilizer recommendations. In *Improving soil fertility recommendations in Africa using the decision support system for agrotechnology transfer (DSSAT)* (pp. 19–42). Springer.

Bommarco, R., Kleijn, D., & Potts, S. G. (2013). Ecological intensification: Harnessing ecosystem services for food security. *Trends in Ecology & Evolution, 28*(4), 230–238.

Boukar, O., Bhattacharjee, R., Fatokun, C., Kumar, P. L., & Gueye, B. (2013) Cowpea. In M. Singh, H. D. Upadhyaya, & I. S. Bisht (Eds.), *Genetic and genomic resources of grain legume improvement* (pp. 137–155).

Choruma, D. J., Akamagwuna, F. C., & Odume, N. O. (2022). Simulating the impacts of climate change on maize yields using EPIC: A case study in the Eastern Cape Province of South Africa. *Agriculture, 12*, 794. https://doi.org/10.3390/agriculture12060794

Choruma, D. J., Balkovic, J., & Odume, O. N. (2019). Calibration and validation of the EPIC model for maize production in the Eastern Cape, South Africa. *Agronomy, 9*(9), 494.

Cole, M. B., Augustin, M. A., Robertson, M. J., & Manners, J. M. (2018). The science of food security. *npj Science of Food, 2*, 14. https://doi.org/10.1038/s41538-018-0021-9

Collins, H. P., Fay, P. A., Kimura, E., Fransen, S., & Himes, A. (2017). Intercropping with switchgrass improves net greenhouse balance in hybrid poplar plantations on a sand soil. *Soil Science Society of America Journal, 81*, 781–795.

De Costa, W. A. J. M. (2011). A review of the possible impacts of climate change on forests in the humid tropics. *Journal of the National Science Foundation of Sri Lanka, 39*, 281–302. https://doi.org/10.4038/jnsfsr.v39i4.3879

Dobermann, A. (2007). Nutrient use efficiency–measurement and management. In *Proceedings of the international fertilizer industry association (IFA) workshop on fertilizer best management practices, 7–9 March 2007, Brussels*, pp. 1–28.

Durr, C., Constantin, J., Wagner, M. H., Navier, H., Demilly, D., Goertz, S., & Nesi, N. (2016). Virtual modeling based on deep phenotyping provides complementary data to field experiments to predict plant emergence in oilseed rape genotypes. *European Journal of Agronomy, 79*, 90–99.

Dwivedi, A., Dev, I., Kumar, V., Yadav, R. S., Yadav, M., Gupta, D., Singh, A., & Tomar, S. S. (2015). Potential role of maize-legume intercropping systems to improve soil fertility status under smallholder farming systems for sustainable agriculture in India. *International Journal of Life Sciences Biotechnology and Pharma Research, 4*(3), 334–340.

Eyshi, R. E., & Bannayan Meteorol, M. (2012). Rainfed wheat yields under climate change in northeastern Iran. *Meteorological Applications, 19*, 346–354. https://doi.org/10.1002/met.268

Eyshi, R. E., Siebert, S., Hüging, H., & Ewert, F. (2018). Climate change effect on wheat phenology depends on cultivar change. *Science and Reports, 2018*(8), 4891. https://doi.org/10.1038/s41598-018-23101-2

FAO. (2003). Food and Agriculture Organization of the United Nations (FAO). Review of world water resources by country. Water Report No. 23; FAO.

FAO. (2012). Coping with water scarcity: An action framework for agriculture and food security. FAO Water Report 38. Rome, Italy.

Ferguson, B. J., Lin, M. H., & Gresshoff, P. M. (2013). Regulation of legume nodulation by acidic growth conditions. *Plant Signal Behavior, 8*(3), e23426.

Fernandez, J. E., & Cuevas, M. V. (2010). Irrigation scheduling from stem diameter variations: A review. *Agricultural and Forest Meteorology, 150*(2), 135–215.

Fghire, R. W., Said, A., Fatima, I. A., Oudou, B. O., & Ragab, R. (2015). Response of quinoa to different water management strategies: Field experiments and SALTMED model application results. *Irrigation and Drainage, 64*(1), 29–40.

Foyer, C. H., Lam, H. M., Nguyen, H. T., et al. (2016). Neglecting legumes has compromised human health and sustainable food production. *Nature Plants, 2*, 16112.

Ghanbari, A., Dahmardeh, M., Siahsar, B. A., & Ramroudi, M. (2010). Effect of maize (Zea mays L.)—Cowpea (Vigna unguiculata L.) intercropping on light distribution, soil temperature and soil moisture in arid environment. *Journal of Food, Agriculture and Environment, 8*(1), 102–108.

Ghosh, P. K., Bandyopadhyay, K. K., Wanjari, R. H., Manna, M. C., Misra, A. K., Mohanty, M., et al. (2007). Legume effect for enhancing productivity and nutrient use-efficiency in major cropping systems–An Indian perspective: A review. *Journal of Sustainable Agriculture, 30*, 59–86. https://doi.org/10.1300/J064v30n01_07

Golabi, M., Naseri, A. A., & Kashkuli, H. A. (2009). Evaluation of SALTMED model performance in irrigation and drainage of sugarcane farms in Khuzestan province of Iran. *Journal of Food, Agriculture and Environment, 7*, 874–880.

Gorissen, A., Tietema, A., Joosten, N. N., Estiarte, M., Peñuelas, J., Sowerby, A., Emmett, B. A., & Beier, C. (2004). Climate change affects carbon allocation to the soil in shrub lands. *Ecosystems, 7*, 650–661.

Hamada, A., & Hamd-Alla, W. (2019). Productivity of intercropped wheat with faba bean under crop sequences and foliar application of humic acid. *Egyptian Journal of Agronomy, 41*(3), 225–241. https://doi.org/10.21608/agro.2019.14396.1167

Hamd-Allah, W. A., Shalaby, E. M., Dawood, R. A., & Zohry, A. A. (2014). Effect of cowpea (Vigan sinensis L.) with maize (Zea mays L.) intercropping on yield and its components. *International Scholarly and Scientific Research & Innovation, 8*(11), 1170–1176.

Harrison, L., Michaelsen, J., Funk, C., & Husak, G. (2011). Effects of temperature changes on maize production in Mozambique. *Climate Research, 46*, 211–222.

Hatfield, J. L., & Prueger, J. H. (2015). Temperature extremes: Effect on plant growth and development. *Weather and Climate Extremes, 10*, 4–10. https://doi.org/10.1016/J.WACE.2015.08.001

Hauggaard-Nielsen, H., Mundus, S., & Jensen, E. S. (2009). Nitrogen dynamics following grain legumes and subsequent catch crops and the effects on succeeding cereal crops. *Nutrient Cycling in Agroecosystems, 84*(3), 281–291. https://doi.org/10.1007/s10705-008-9242-7

Hawkesford, M. J., Araus, J. L., Park, R., Calderini, D., Miralles, D., Shen, T., Zhang, J., & Parry, M. A. J. (2013). Prospects of doubling global wheat yields. *Food Energy Security, 2*, 34–48.

Hernandez-Ochoa, M. I., Asseng, S., Kassie, B., Xiong, W., Robertson, R., Diego Notelo Luz Pequeno, D. N. L., Sonder, K., Reynolds, M., Babar, M. A., Milan, A. M., & Hoogenboom, G. (2018). Climate change impact on Mexico wheat production. *Agricultural and Forest Meteorology,* (263), 373–387.

Hirich, A., Ragab, R., Jacobsen, S. E., El Yousfi, L., & El Omari, H. (2012). Using deficit irrigation with treated wastewater in the production of quinoa (Chenopodium quinoa Willd.) in Morocco. *Revista Cientıfica UDO Agrıcola, 12*(3), 570–583.

Hussain, J., Khaliq, T., Ahmad, A., Akhter, J., & Asseng, S. (2018). Wheat responses to climate change and its adaptations: A focus on arid and semi-arid environment. *International Journal of Environmental Research, 12*, 117–126. https://doi.org/10.1007/s41742-018-0074-2

IPCC. (2007). *Contribution of working group II to the fourth assessment report of the intergovernmental panel on climate variability* (p. 60). Cambridge University Press.

Ivanov, S., Fedorova, E. E., Limpens, E., De Mita, S., Genre, A., Bonfante, P., et al. (2012). Rhizobium–legume symbiosis shares an exocytotic pathway required for arbuscule formation. *Proceedings of the National Academy of Sciences, 109*, 8316–8321. https://doi.org/10.1073/pnas.1200407109

Jamieson, P. D., Porter, J. R., Goudriaan, J., Ritchie, J. T., Keulen, H., & Stol, W. (1998). A comparison of the models AFRCWHEAT2, CERES-Wheat, Sirius, SUCROS2 and SWHEAT with measurements from wheat grown under drought. *Field Crops Research, 55*, 23–44.

Kadiyala, M. D. M., Nedumaran, S., Singh, P., Chukka, S., Irshad, M. A., & Bantilan, M. C. S. (2015). An integrated crop model and GIS decision support system for assisting agronomic decision making under climate change. *Science of the Total Environment, 521*, 123–134.

Kariaga, B. M. (2004). Intercropping maize with cowpeas and beans for soil and water management in Western Kenya. In *Proceedings 13th international soil conservation organization conference, conserving soil and water for society*, Brisbane, pp. 1–5.

Karmakar, R., Das, I., Dutta, D., & Rakshit, A. (2016). Potential effects of climate change on soil properties: A review. *Science International, 4*, 51–73.

Kebede, E. (2020). Grain legumes production and productivity in Ethiopian smallholder agricultural system, contribution to livelihoods and the way forward. *Cogent Food Agriculture, 6*, 1722353. https://doi.org/10.1080/23311932.2020.1722353

Knox, J., Daccache, A., Hess, T., & Haro, D. (2016). Meta-analysis of climate impacts and uncertainty on crop yields in Europe. *Environmental Research Letters, 11*, 113004.

Kogo, B. K., Kumar, L., Koech, R., & Langat, P. (2019). Modelling impacts of climate change on maize (Zea mays L.) growth and productivity: A review of models, outputs and limitations. *Journal of Geoscience and Environmental Protection, 7*, 76–95.

Lemage, B., & Tsegaye, M. (2020). Evaluation of legume shrubs improved fallow for abandoned agricultural land rehabilitation. *International Journal of Agricultural Research, Innovation and Technology, 10*, 64–70. https://doi.org/10.3329/ijarit.v10i1.48095

Lemke, R. L., Zhong, Z., Campbell, C. A., & Zentner, R. P. (2007). Can pulse crops play a role in mitigating greenhouse gases from North American agriculture? *Agronomy Journal, 99*, 1719–1725.

Li, C. J., Hoffland, E., Kuyper, T. W., Yu, Y., Li, H., Zhang, C., Zhang, F., & van der Werf, W. (2020). Yield gain, complementarity and competitive dominance in intercropping in China: A meta-analysis of drivers of yield gain using additive partitioning. *European Journal of Agronomy, 113*, 125987. https://doi.org/10.1016/j.eja.2019.125987

Li, C. S., Tang, Y. L., Wu, C., & Huang, G. (2012). Effect of sowing patterns on growth, development and yield formation of wheat in rice stubble land. *Transactions of CSAE, 28*, 36–43.

Lithourgidis, A. S., & Dordas, C. A. (2010). Forage yield, growth rate, and nitrogen uptake of faba bean intercrops with wheat, barley, and rye in three seeding ratios. *Crop Science, 50*, 2148–2158. https://doi.org/10.2135/cropsci2009.12.0735

Liu, B., Asseng, S., Muller, C., Ewert, F., & Elliott, J. (2016). Similar estimates of temperature impacts on global wheat yield by three independent methods. *Nature Climate Change*, 1130–1136.

Liu, J., & Yang, H. (2010). Spatially explicit assessment of global consumptive water uses in cropland: Green and blue water. *Journal of Hydrology, 384*, 187–197.

Lizaso, J., Ruiz-Ramos, M., Rodríguez, L., Gabaldon-Leal, C., Oliveira, J., Lorite, I., Sánchez, D., García, E., & Rodríguez, A. (2018). Impact of high temperatures in maize: Phenology and yield components. *Field Crops Research, 216*, 129–140. https://doi.org/10.1016/j.fcr.2017.11.013

Lopez-Bellido, L., Lopez-Bellido, R. J., Redondo, R., & Benitez, J. (2006). Faba beans nitrogen fixation in a wheat-based rotation under rain fed Mediterranean conditions: Effect of tillage system. *Field Crops Research, 98*, 253. https://doi.org/10.1016/j.fcr.2006.03.001

Masaka, J. (2006). The effect of n fertilizer placement and timing on the aboveground biometric characteristics of spring wheat (Triticum aestivum L. cv. Spectrum) on Leached Chernozem. *International Journal of Agricultural Research, 1*, 68–75.

Mayer, J., Buegger, F., Jensen, E. S., Schloter, M., & Hess, J. (2003). Estimating N rhizodeposition of grain legumes using a 15N in situ stem labeling method. *Soil Biology & Biochemistry, 35*, 21–28. https://doi.org/10.1016/S0038-0717(02)00212-2

McCallum, M. H., Kirkegaard, J. A., Green, T., Cresswell, H. P., Davies, S. L., & Angus, J. F. (2004). Improved subsoil macro-porosity following perennial pastures. *Australian Journal of Experimental Agriculture, 44*, 299–307.

Moneir, I. (2012). *Simulation of the effect of adaptation strategies on improving yield of some crops grown under expected climate change conditions* [MSc Thesis, Institute of Environmental Research Studies, Ain Shams University, Egypt].

Morsy, M. (2015). *Use of regional climate and crop simulation models to predict wheat and maize productivity and their adaptation under climate change* [PhD thesis, Faculty of Science Al-Azhar University].

Msowoya, K., Madani, K., Davtalab, R., Mirchi, A., & Lund, J. R. (2016). Climate change impacts on maize production in the warm heart of Africa. *Water Resources Management, 30*, 5299–5312. https://doi.org/10.1007/s11269-016-1487-3

Neiff, N., Trachsel, S., Valentinuz, O. R., Balbi, C. N., & Andrade, F. H. (2016). High temperatures around flowering in maize: Effects on photosynthesis and grain yield in three genotypes. *Crop Science, 56*, 2702–2712. https://doi.org/10.2135/cropsci2015.12.0755

Noreldin, T., Abdrabbo, M., & Ouda, S. (2012). Increasing water productivity for wheat grown under climate change conditions. In *10th international conference of Egyptian soil science society (ESSS) and 4th international conference of water requirements & metrology Department, 5–7 November*, Ameria, Egypt (Cl 06).

Nuruzzaman, M., Lambers, H., Bolland, M. D. A., & Veneklaas, E. J. (2005). Phosphorus benefits of different legume crops to subsequent wheat grown in different soils of Western Australia. *Plant and Soil, 271*, 175–187. https://doi.org/10.1007/s11104-004-2386-6

Ooro, P. A., Malinga, J. N., Tanner, D. G., & Payne, T. S. (2011). Implication of rate and time of nitrogen application on wheat (Triticum aestivum L.) yield and quality in Kenya. *Journal of Animal and Plant Sciences, 9*(2), 1141–1146.

Osborne, C. A., Peoples, M. B., & Janssen, P. H. (2010). Detection of a reproducible, single-member shift in soil bacterial communities exposed to low levels of hydrogen. *Applied Environmental Microbiology, 76*, 1471–1479.

Ouda S., & Zohry, A. A. (2017). Crops intensification to reduce wheat gap in Egypt. In *Future of food gaps in Egypt: Obstacles and opportunities*. Springer Publishing House. ISBN 978-3-319-46942-3.

Ouda, S., Abdrabbo, M., & Noreldin, T. (2012a). Effect of changing sowing dates and irrigation scheduling on maize yield grown under climate change conditions. In *10th international conference of Egyptian soil science society (ESSS) and 4th international conference of water requirements & metrology department, 5–7 November*, Ameria, Egypt (Cl 07).

Ouda, S., Noreldin, T., AbouElenin, R., & Abd, E.-B. (2012b). Improved agricultural management practices reduced wheat vulnerably to climate change in salt affected soils. *Egyptian Journal of Agricultural Research, 904*, 499–513.

Ouda S., Zohry, A., & Abdel-Wahab, S. (2020). Sustainable intensive cropping to reduce irrigation-induced erosion: II. Changing cropping sequence under sprinkler irrigation practice. *Moroccan Journal of Agricultural Science, 1*(1), 26–33.

Ouda, S. A., Khalil, F. A., & Yousef,H. (2009). Using adaptation strategies to increase water use efficiency for maize under climate change conditions. In *Proceeding of 13th international conference on water technology, Egypt.*

Ouda, S. A., Sayed, M., El Afandi, G., & Khalil, F. A. (2010). Developing an adaptation strategy to reduce climate change risks on wheat grown in sandy soil in Egypt. In *Proceeding of 10th international conference on development of dry lands, Egypt.*

Parry, M. A. J., Reynolds, M., Salvucci, M. E., Raines, C., Andralojc, P. J., Zhu, X.-G., Price, G. D., Condon, A. G., & Furbank, R. T. (2011). Raising yield potential of wheat. II. Increasing photosynthetic capacity and efficiency. *Journal of Experimental Botany, 62*, 453–467.

Pask, A., Joshi, A., Manes, Y., Sharma, I., Chatrath, R., Singh, G., Sohu, V., Mavi, G., Sakuru, V., & Kalappanavar, I. (2014). A wheat phenotyping network to incorporate physiological traits for climate change in South Asia. *Field Crops Research, 168*, 156–167. https://doi.org/10.1016/j.fcr.2014.07.004

Patricia, O., Yeboah, S., Addy, S. N., Amponsah, S., & Owusu, D. E. (2013). Crop modeling: A tool for agricultural research–a review. *Journal of Agricultural Research and Development, 2*(1), 1–6.

Paymard, P., Bannayan, M., & Haghighi, R. S. (2018). Analysis of the climate change effect on wheat production systems and investigate the potential of management strategies. *Natural Hazards, 91*, 1237. https://doi.org/10.1007/s11069-018-3180-8

Porter, J. R., & Semenov, M. A. (2005). Crop responses to climatic variation. *Philosophical Transactions of the Royal Society of London. Series b, Biological Sciences, 360*, 2021–2035.

Pulvento, C., Lavini, A., Riccardi, M., d'Andria, R., & Ragab, R. (2015). Assessing amaranth adaptability in a Mediterranean area of south Italy under different climatic scenarios. *Irrigation and Drainage, 64*(1), 50–58.

Pulvento, C., Riccardi, M., Lavini, A., D'andria, R., & Ragab, R. (2013). SALTMED model to simulate yield and dry matter for quinoa crop and soil moisture content under different irrigation strategies in south Italy. *Irrigation and Drainage, 62*(2), 229–238.

Pypers, P., Huybrighs, M., Diels, J., Abaidoo, R., Smolders, E., & Merckx, R. (2007) Does the enhanced P acquisition by maize following legumes in a rotation result from improved soil P availability? *Soil Biology and Biochemistry, 39*, 2555. https://hdl.handle.net/10568/92833

Ragab, R. (2002). A holistic generic integrated approach for irrigation, crop and field management: The SALTMED model. *Environmental Modelling & Software, 17*(4), 345–361.

Ragab, R. (2015). Integrated management tool for water, crop, soil and N-fertilizers: The saltmed model. *Irrigation and Drainage, 64*(1), 1–12.

Ragab, R., Malash, N., Gawad, G. A., Arslan, A., & Ghaibeh, A. (2005a). A holistic generic integrated approach for irrigation, crop and field management: 1. The SALTMED model and its calibration using field data from Egypt and Syria. *Agricultural Water Management, 78*(1), 67–88.

Ragab, R., Malash, N., Gawad, G. A., Arslan, A., & Ghaibeh, A. (2005b). A holistic generic integrated approach for irrigation, crop and field management: 2. The SALTMED model validation using field data of five growing seasons from Egypt and Syria. *Agricultural Water Management, 78*(1), 89–107.

Razzaghi, F., Plauborg, F., Ahmadi, S. H., Jacobsen, S.-E., Andersen, M. N., & Ragab, R. (2011). Simulation of quinoa (Chenopodium quinoa Willd.) response to soil salinity using the SALTMED model. In *ICID 21st International congress on irrigation and drainage, 15–23 October 2011*, International Commission on Irrigation and Drainage, Tehran, Iran, pp. 25–32.

Rötter, R. P., Carter, T. R., Olesen, J. E., & Porter, J. R. (2011). Crop–climate models need an overhaul. *Nature Climate Change, 1*(4), 175–177.

Ruisi, P., Amato, G., Badagliacca, G., Frenda, A. S., Giambalvo, D., & Di Miceli, G. (2017) Agroecological benefits of faba-beans for rainfed Mediterranean cropping systems. *Italian Journal of Agronomy, 12*(3), 223. https://doi.org/10.4081/ija.2017.865

Sánchez, B., Rasmussen, A., & Porter, J. R. (2014). Temperatures and the growth and development of maize and rice: A review. *Global Change Biology, 20*, 408–417. https://doi.org/10.1111/gcb.12389

Schauberger, B., Archontoulis, S., Arneth, A., Balkovic, J., Ciais, P., Deryng, D., et al. (2017). Consistent negative response of US crops to high temperatures in observations and crop models. *Nature Communications, 8*(1), 1–9. https://doi.org/10.1038/ncomms13931

Scott, R. K., English, S. D., Wood, D. W., & Unsworth, M. H. (2009). The yield of sugar beet in relation to weather and length of growing season. *The Journal of Agricultural Science, Cambridge, 81*, 339–347.

Sheha, A. M., Nagwa, R., & Ahmed, A.-E. (2014). Effect of crop sequence and nitrogen levels on rice productivity. *Annals of Agricultural Science, 52*(4), 451–460.

Shim, D., Lee, K.-J., & Lee, B.-W. (2017). Response of phenology-and yield-related traits of maize to elevated temperature in a temperate region. *Crop Journal, 5*, 305–316. https://doi.org/10.1016/j.cj.2017.01.004

Silva, L. L., Ragab, R., Durante, I., Lourenço, E., Simões, N., & Chaves, M. M. (2013). Calibration and validation of SALTMED model under dry and wet conditions using chickpea field data from southern Portugal. *Irrigation Science, 31*, 651–659.

Singh, B. B. (2014). *Cowpea: The food legume of the 21st century*. Crop Science Society of America.

Singh, B. B., Ajeigbe, H. A., Tarawali, S. A., Fernandez-Rivera, S., & Musa, A. (2003). Improving the production and utilization of cowpea as food and fodder. *Field Crops Research, 84*, 169–177.

Stagnari, F., Maggio, A., Galieni, A., & Pisante, M. (2017). Multiple benefits of legumes for agriculture sustainability: An overview. *Chemical and Biological Technologies in Agriculture, 4*, 1–13. https://doi.org/10.1186/s40538-016-0085-1

Steduto, P., Raes, D., Hsiao, T., Fereres, E., Heng, L., Howell, T., & Geerts, S. (2009). Concepts and applications of AquaCrop: The FAO crop water productivity model. In W. Cao, J. White, & E. Wang (Eds.), *Crop modeling and decision support* (pp. 175–191). Springer.

Taha, A., & Ouda, S. (2016). Deficit irrigation for wheat and maize grown in sandy soil to face water scarcity. In *4th African regional ICID conference, 24–28 April*.

Teixeira, E. I., Fischer, G., Velthuizen, H. V., Walter, C., & Ewert, F. (2013). Global hot spots of heat stress on agricultural crops due to climate change. *Agricultural and Forest Meteorology, 170*, 206–215. https://doi.org/10.1016/j.agrformet.2011.09.002

Thilakarathna, M. S., McElroy, M. S., Chapagain, T., Papadopoulos, Y. A., & Raizada, M. N. (2016). Belowground nitrogen transfer from legumes to non-legumes under managed herbaceous cropping systems. A review. *Agronomy for Sustainable Development, 368*. https://doi.org/10.1007/s13593-016-0403-9

UNEP. (2019). *Frontiers 2018/19: Emerging issues of environmental concern*. UNEP.

Valipour, M. (2014). Pressure on renewable water resources by irrigation to 2060. *Acta Advances in Agricultural Sciences, 2*, 32–42. http://www.aaasjournal.org

Valizadeh, J., Ziaei, S. M., & Mazloumzadeh, S. M. (2014). Assessing climate change impacts on wheat production (a case study). *Journal of the Saudi Society of Agricultural Sciences, 13*(2), 107–115. https://doi.org/10.1016/j.jssas.2013.02.002

Vasconcelos, M. W., Grusak, M. A., Pinto, E., Gomes, A., Ferreira, H., Balázs, B., et al. (2020). The biology of legumes and their agronomic, economic, and social impact. In *The plant family fabaceae* (pp. 3–25). Springer. https://doi.org/10.1007/978-981-15-4752-2_1

Vijayalakshmi, K., Fritz, A. K., Paulsen, G. M., Bai, G., Pandravada, S., & Gill, B. S. (2010). Modeling and mapping QTL for senescence-related traits in winter wheat under high temperature. *Molecular Breeding, 26*, 163–175.

Wan, Y., Lin, E., Xiong, W., Li, Y., & Guo, L. (2011). Modeling the impact of climate change on soil organic carbon stock in upland soils in the 21st century in China. *Agriculture, Ecosystems & Environment, 141*, 23–31.

Wang, L., Guo, L., Li, Y., Wang, Z., et al. (2012). Systematic comparison of C3 and C4 plants based on metabolic network analysis chuanli BMC. *Systems Biology, 6*(Suppl 2), S9. http://www.biomedcentral.com/1752-0509/6/S2/S9

Wezel, A., Casagrande, M., Celette, F., Vian, J. F., Ferrer, A., & Peigné, J. (2014). Agroecological practices for sustainable agriculture. *Agronomy for Sustainable Development, 34*(1), 1–20.

White, J. W., & Reynolds, M. P. (2003). A physiological perspective on modeling temperature response in wheat and maize crops. In J. W. White (Ed.), *Modeling temperature response in wheat and maize. Proceedings of a workshop, CIMMYT*, El Batán, Mexico, 23–25 April 2001. CIMMYT, Mexico City, pp. 8–17.

Willmott, C. J. (1981). On the validation of models. *Physical Geography, 2*, 184–194.

Wollenberg, E., Richards, M., Smith, P., Havlík, P., Obersteiner, M., Tubiello, F. N., Herold, M., Gerber, P., Carter, S., Reisinger, A., van Vuuren, D. P., Dickie, A., Neufeldt, H., Sander, B. O., Wassmann, R., Sommer, R., Amonette, J. E., Falcucci, A., Herrero, M., …, Campbell, B. M. (2016). Reducing emissions from agriculture to meet the 2 °C target. *Global Change Biology, 22*(12), 3859–3864. https://doi.org/10.1111/gcb.13340. Epub, Jul 11.

World Bank Group. (2021). Climate change knowledge portal (CCKP) user manual (18 p). World Bank Group.

Wu, W., Fan, L., Li, M. F., Liu, H. B., & Li, Y. Q. (2012). Sensitivity analysis of crop growth models to multi-temporal scale solar radiation. *Transactions of the Chinese Society of Agricultural Engineering, 28*, 123–128.

Yang, S., Logan, J., & Coffey, D. L. (1995). Mathematical formulae for calculating the base temperature for growing degree days. *Agricultural and Forest Meteorology, 74*(1–2), 61–74. https://doi.org/10.1016/0168-1923(94)02185-M

Yusuf, A. A., Iwuafor, E. N. O., Abaidoo, R. C., Olufajo, O. O., & Sanginga, N. (2009). Grain legume rotation benefits to maize in the northern Guinea savanna of Nigeria: Fixed-nitrogen versus other rotation effects. *Nutrient Cycling in Agroecosystems, 84*, 129. https://doi.org/10.1007/s10705-008-9232-9

Zafar, J., & Muhammad, F. C. (2007). Effects of soil and foliar application of different concentrations of NPK and foliar application of $(NH_4)_2$ SO_4 on growth and yield attributes in wheat (Triticum aestivum L.). *Pakistan Journal of Plant Science, 13*(2), 119–128.

Zhao, C., Liu, B., Piao, S., Wang, X., Lobell, D. B., Huang, Y., Huang, M., Yao, Y., Bassu, S., Ciais, P., Durand, J.-L., Elliott, J., Ewert, F., Janssens, I. A., Li, T., Lin, E., Liu, Q., Martre, P., Müller, C., …, Asseng, S. (2017). Temperature increase reduces global yields of major crops in four independent estimates. *PNAS, 114*(35), 9326–9331. https://doi.org/10.1073/pnas.1701762114

Zohry, A. A. (2005). Effect of preceding winter crops and intercropping on yield, yield components and associated weeds in maize. *Annals of Agricultural Science, Moshtohor, 43*(1), 139–148.

Zohry, A. A., Ouda, S., Hamd-Alla, W., & Shalaby, E. (2017). Evaluation of different crop sequences for wheat and maize in sandy soil. *Acta Agriculturae Slovenica, 109*(2), 383–392.

Chapter 5
Assessment of the Impact of Climate Change on Rice Productivity: Modeling and Simulation Studies

Abstract Rice is the third most important cereal globally after maize and wheat. It is the main food crop for more than 60% of the human population and it occupies almost one-fifth of the total land area covered under cereals. There are two types of rice systems, namely paddy and upland rice. Paddy rice requires high irrigation amounts for proper growth and development and consequently its water use efficiency is low. On the other hand, upland rice has strong drought resistance and low water requirements. In this chapter, we used a modeling approach to assess the impact of future climate change on water requirements and water productivity of three cultivars of paddy rice planted in a two-year field experiment in Egypt using future climate scenarios developed by CMIP6, namely SSP1-2.6 and SSP5-8.5 in two time-intervals, namely 2060–2079 and 2080–2099. We also employed a simulation approach using SALTMED model to assess the impact of climate change under the above scenarios on the yield of upland rice grown under deficit irrigation. Furthermore, we reviewed the attempts to intercrop legume crops with upland rice to increase its productivity in a sustainable manner. The results indicated that the required irrigation water will increase under the two future scenarios, where the increase was the highest under SSP5-8.5 in 2080–2099. Consequently, water productivity will be also reduced under the studied future scenarios. Moreover, a two-year field experiment was implemented to assess the impact of climate change on upland rice grown under deficit irrigation treatments. The climate-suitability of upland rice cultivation in the future was tested by graphing the cutoff temperature (T_{cut}, 35 °C) with maximum temperature (T_{max}) under the studied future scenarios to determine the number of days that T_{cut} will exceed T_{max}. The results showed that the frequency of T_{cut}, exceeding T_{max} was the lowest under SSP1-2.6 in 2060–2079 and it was the highest under SSP5-8.5 in 2080–2099, which implied higher yield losses. An increase in the water requirements of upland rice was projected under the studied future scenarios. Using SALTMED model to project the yield of upland rice revealed that, under deficit irrigation, yield losses were higher than the values projected under full irrigation application and was the highest under SSP5-8.5 in 2080–2099. The above results suggest that the cultivation of upland rice will be more suitable to conserve water resources in the future. Furthermore, intercropping upland rice with legume crops, namely peanut, mung bean, and soybean proved to

S. Ouda and A. E. Zohry, *Integration of Legume Crops with Cereal Crops Under Changing Climate*, https://doi.org/10.1007/978-3-031-68102-8_5

be effective in increasing the yield of upland rice and it could be implemented in the future to reduce the harmful effect of climate change on rice production.

Keywords Paddy rice · Upland rice · Heat stress · Climate change · CMIP6 · Deficit irrigation · Climate-suitability · Water productivity · SALTMED model

Introduction

Rice is the third most important cereal globally after maize and wheat (Yadav et al., 2019). It is the main food crop for more than 60% of the human population (Joshi & Karan, 2014), and it occupies almost one-fifth of the total land area covered under cereals (Anis et al., 2019). It is expected that in 2050 the demand for rice will increase by 70% (FAO, 2018).

There are two types of rice production systems, namely paddy rice and upland rice. Paddy rice is a semi aquatic crop that requires high irrigation amounts for proper growth and development and consequently its water use efficiency is low (Wassman, 2019). Its cultivation has been practiced in lowland regions of Asia for hundreds of years (Bouman & Tuong, 2001). Bouman et al. (2007) indicated that paddy rice cultivation is very water-intensive and accounts for 24–30% of global freshwater consumption alone. Several researchers made efforts in applying techniques to save in the applied the irrigation water to paddy rice (Wang et al., 2016). An example of these technologies is alternating wetting and drying, which was found to have a little negative effect on yield of paddy rice (Chu et al., 2015). This technique could save about 15–30% of water use without any yield reduction (Runkle et al., 2019). Kumar et al. (2017) stated that mild alternating wetting and drying saved about 23% of the applied water with no decrease in yield, where severe alternating wetting and drying significantly reduce rice grain yield in comparison with continued flooding. Furthermore, furrow irrigation was found to be more efficient than conventional flooded irrigation in water conservation, as well as obtaining reasonable yield (Carroll et al., 2020). Zhang et al. (2011) reported 20–50% saving in water used for irrigating paddy rice planted on furrows, since water is only applied to furrows, which reduces surface water area, evaporation, seepage and deep percolation, compared to conventional flooded irrigation (Rai et al., 2017). Zeng and Li (2020) found an increase in paddy rice grain yield under furrow irrigation by 11–18%, compared to the conventional flooded irrigation.

Moreover, emission of CH_4 and N_2O form paddy rice fields have been a major problem in global warming situation. Several field managements were studied to reduced emission of CH_4 and N_2O form paddy rice fields. Hadi et al. (2010) indicated that intermittent drainage could reduce greenhouse gases emission by from 15 to 67%, compared to continuous flooding. Whereas, alternate wetting and drying could reduce the emissions by 33–39%, compared to continuous flooding with no yield differences (Setyanto et al., 2017). Intermittent irrigation and saturated soil have less CH_4 emission, compared to continuous flooding (Husin et al., 1995). Additionally, applications of fresh and decomposed straw to paddy rice fields could reduce N_2O emission by around 33 and 28% compare to without straw application (Wihardjaka et al., 2010). Water savings techniques applied to paddy rice field can largely contribute to reduction of these two gases (Djaman et al., 2018).

Conversely, upland rice has strong drought resistance and lower water requirements than the paddy rice, thus it is suitable to be cultivated in arid and low mountain areas (Marschner, 1978). It can be used as a substitute for paddy rice cultivation, where rainfall is stable but irrigation conditions are lacking (Kukal et al., 2005). Leng and Hall (2019) reported that the upland rice cultivars are able to grow without standing water in aerobic soils, make upland rice similar to wheat or maize irrigation. A large number of upland rice varieties have been developed by breeding programs and their yield and quality are approaching those of paddy rice varieties (Atlin et al., 2013). The cultivation of the upland rice varieties can play an important role stabilizing and promoting food security (Yang et al., 2007), as well as improving the utilization of water resources (Wang et al., 2016), and reducing environmental problems caused by flooded paddy rice cultivation (Christy et al., 2018). The plants of upland rice are seeded directly without continuous flooded conditions, which is a promising approach to save a remarkable quantity of irrigation water. Fukai and Mitchell (2022) reported that the applied irrigation water to upland rice was lower by 44%, compared to traditional transplanted systems.

The current situation of the prevailing water scarcity worldwide encourages the use of deficit irrigation in irrigating crops, where it has been considered as a key contributor in water saving technology on field level (Capra et al., 2008). Deficit irrigation application can be very useful in reducing the demand for water in the agriculture sector (Fereres & Soriano, 2007). Deficit irrigation was defined by Chai et al. (2016) as "an irrigation practice characterized by application of irrigation water below the full required amounts for optimal growth and yield, aiming at improving the response of the plants to the certain degree of water deficit in a positive manner, and improving crop's water use efficiency". Additionally, Kögler and Söffker (2017) indicated that deficit irrigation is a crop cultivation practice allows saving up to 20–40% of irrigation water with yield reductions below 10%, leading to more efficient water use.

The effect of water deficiency application on upland rice have been extensively studied worldwide (Akinbile, 2010; Crusciol et al., 2013; Dingkuhn et al., 2015; Kato et al., 2006a, 2006b, 2007). In Egypt, application of deficit irrigation to upland rice, namely increasing irrigation intervals, was studied by few authors. Sheha et al. (2021) found that increasing irrigation interval from 6 to 9 days, reduced the yield of upland rice by 12%, whereas the reduction was 36% under irrigation every 12 days, compared to irrigation every 6 days. Furthermore, Abd El-Naby and El-Ghandor (2022) reported a decrease in the yield of upland rice by 21 and 50% under irrigation every 8 and 12 days, respectively, compared to irrigation every 4 days.

Climate change poses as a huge challenge to food security, as it will impact all food security components and the entire food system (Vermeulen et al., 2012). Climate models in the 5th phase of Coupled Model Intercomparison Project (CMIP5) are used to simulate and project the effect of climate change by climate related scenarios called representative concentration pathways (RCPs) (N'Guessan et al., 2023). RCPs provide climate projections on short-time scale of future, namely up to 2035 and long-time scale of future, namely up to 2100, thus facilitate understanding of differences in climate model projections (Bauwens et al., 2013). Whereas, in the 6th phase of Coupled Model Intercomparison Project (CMIP6), a new generation of GCMs has been developed (Eyring et al., 2016). One novelty is the "Shared Socio-economic Pathways" are combined with the CMIP5 GCMs. These combinations aim to develop future climate change conditions by ensuring major improvements in emissions, land use scenarios, improved models for parameterization and physical process, among other benefits (Eyring et al., 2016).

In this chapter, we used a modeling approach to assess the impact of future climate change on water requirements and water productivity of three cultivars of paddy rice planted in a two-year field experiment in Egypt using future climate scenarios developed in CMIP6, namely SSP1-2.6 and SSP5-8.5 in two time-intervals, namely 2060–2079 and 2080–2099. We also employed a simulation approach using SALTMED model to assess the impact of climate change on the yield of upland rice grown under deficit irrigation. Furthermore, we reviewed the attempts to intercrop legume crops with upland rice to increase its productivity in a sustainable manner.

Effect of Heat Stress on Rice Growth and Yield

A negative effect of heat stress has been reported on rice growth and yield. The most susceptible growth stage to heat stress in rice is flowering stage, where temperatures of 32–36 °C cause high spikelet sterility (Satake & Yoshida, 1978), which is associated with grain yield reduction. High temperatures above 35 °C at the flowering stage inhibit anther growth and causing incomplete fertilization (Jagadish et al., 2007). High temperature stress during rice flowering stage led to decline in pollen number and shed (Prasad et al., 2006). Mahmood et al. (2010) indicated that exposing rice plants to heat waves during flowering growth stage could cause sterility, whereas during grain filling stage could reduce grain weight and grain number. Heat stress occurred during the reproductive organ developing resulted in severe loss in rice productivity (Zou et al., 2009). Otengdarko et al. (2012) reported a decline in rice yield by 60% due to 4 °C increase in temperature. Photosynthesis of rice was reduced by 11–36%, when it was exposed to 3.6 and 7.0 °C higher temperature than ambient from heading to middle ripening stage, respectively (Oh-e et al., 2007). Temperature higher than 35 °C for more than 3 days during reproductive stages can affect pollen development and pollination, resulting in decreased seed setting and production (Welch et al., 2010).

Effect of Climate Change on Rice Productivity

The risk of high temperature stress under climate change would notably increase the losses in crops production (Fulu & Zhang, 2013), where sustainable production of rice is mainly challenged by climate change (Hina et al., 2019). Changes in the climate system have had a general impact on global food production, and the risk of climate change severely affecting crop yields in the future may also increase (Pachauri & Meyer, 2014). Assessing the impact of climate change on agricultural production systems is mainly done using crop models associated with climate model outputs (N'guessan et al., 2023).

The effect of climate change on rice has been studied by several researchers. Nasir et al. (2020) reported a decrease in rice grain yield will occur during future scenario due to climate warming. Using CMIP6 for climate change study, Elsadek et al., 2024 reported a decline in rice yield by 7 and 9% under SSP2–4.5 future scenario and by 23 and 34% under SSP5–8.5 future scenario during the 2070s and 2090s, respectively. Sharma et al. (2022) reported a decline in rice yield in the US by 8% for a 1 °C rise in maximum temperature under future climate change. Ceccarelli et al. (2010) reported losses in rice yields up to 41% by the end of the 21st Century under future climate change. Whereas, in China, Yin et al. (2015) indicated that the yield of rice is expected to decline without CO_2 fertilization effect and it will increase with CO_2 fertilization effect. Furthermore, In India and under two climate change models, i.e. REMO, and HadRM3 in 2040, Mishra et al. (2013) projected a decrease in rice yield by 25%. Additionally, Kim et al. (2011), in Korea, projected significant yield decreases in rice by 22 and 35% in 2050 and 2100, respectively as a result of future climate change.

Paddy Rice Cultivation in Egypt

Rice is considered the second important food crop after wheat and it is a main food crop for more than half of the world population. In Egypt, rice is the third vital cereal crop after wheat and maize, which represents an essential part in Egyptian food basket. In 2020, the total cultivated area of rice was 495.2 thousand hectares, which produced 4.4 million ton of rice, with an average productivity of 8.97 ton/ha (Central Administration for Agricultural Economics, 2020). Rice is mainly cultivated in the Nile Delta of Egypt to protect its lands from sea water intrusion. The conventional irrigation system for rice is flooding and maintaining free water in the field. Continual flooding provides a favorable water and nutrient supply under the anaerobic conditions (Mahmoud et al., 2016). However, this conventional system involves high water loss through evaporation, seepage and percolation (Abd El-Naby & El-Ghandor, 2022).

Modeling the Impact of Climate Change on Paddy Rice Cultivated in North Egypt

We used published data on paddy rice by Noreldin and Abd El-Bakey (2017) from a two-year field experiment. The field experiment was implemented at Demiatte Governorate (31° 07 N, 30° 57 E), during 2013 and 2014 summer growing seasons. Three rice cultivars, namely Giza177, Giza178, and Sakha101 were grown under submerge conditions with continuous flooding to maintain 7 cm of water depth. Planting date was 8th and 9th of May in the first and second growing seasons, respectively. Season length was different for the three cultivars, namely 125, 135, 140 days for Giza177, Giza178, and Sakha101, respectively. The three rice cultivars were irrigated every 6 days, which is the recommended irrigation practice and it was irrigated every 12 days as a deficit irrigation practice.

The soil of the experiment was clayey in texture, with 65.95% clay, 11.79% sand and 22.26% silt. The soil field capacity, wilting point and available water were 45.96, 24.97, and 25.84% and bulk density value was 1.16 g/cm^3.

The Yield, the Applied Irrigation Water and Water Productivity of the Paddy Rice Cultivars

Table 5.1 presents the yield and the applied irrigation water to the three rice cultivars, under the conditions of the field experiment. The results indicated that irrigation every 12 days reduced rice yield by 32, 22, and 29% for Giza177, Giza178 and Sakha101, respectively and contributed in 14, 16 and 15% saving in the applied irrigation water, respectively compared to irrigation every 6 days. The losses in the yield of the three paddy rice cultivars can be explain by the findings of Vibhuti et al. (2015), where they indicated that low water availability in the early seedling stage of paddy rice leads to unbalanced and poor stand seedlings establishment. Low water availability also interrupt active seed germination, causing osmotic imbalance, membrane impairment, decreased respiration, and ATP production (Kadam et al., 2017). Furthermore, during the vegetative period, water stress causes a delayed in panicle initiation, induces late maturity (Singh et al., 2012), which associated with decline in the yield. Whereas, water stress during the reproductive growth stage was found to be the most damaging on rice grain yield (Alam Khan, 2012).

Table 5.1 also presents the values of water productivity obtained under the two irrigation treatments. Water productivity is a quantitative term used to define the relationship between crop produced and the amount of water involved in crop production (Chai et al., 2016). The results showed that the values of water productivity were negatively affected by water deficiency (irrigation every 12 days). The highest water productivity values were obtained for Giza 178, namely 0.59 and 0.55 kg/m^3 under irrigation every 6 and 12 days, respectively. The average water productivity values

Table 5.1 The yield of paddy rice cultivars, the applied irrigation water and water productivity (WP) averaged over the two growing seasons

	Irrigation every 6 days			Irrigation every 12 days		
	Yield (ton/ha)	Irrigation (m^3/ha)	WP (kg/m^3)	Yield (ton/ha)	Irrigation (m^3/ha)	WP (kg/m^3)
Giza177	7.44	13,166	0.57	5.04	11,261	0.45
Giza178	8.64	14,530	0.59	6.72	12,271	0.55
Sakha101	8.40	14,892	0.56	6.00	12,595	0.48
Average	8.16	14,196	0.57	5.92	12,042	0.49

Source Noreldin and Abd El-Bakey (2017)

over the three cultivars were 0.57 and 0.49 kg/m^3 under irrigation every 6 and 12 days, respectively.

Projection of the Water Requirements of Paddy Rice Using CMIP6 Future Scenarios

To assessed the effect of climate change on the required irrigation water of the three paddy rice cultivars, the CMIP6 future scenarios were used. The data of CMIP6 mean projection of two future climate scenarios, namely SSP1-2.6 and SSP5-8.5 were collected from The World Bank Group, Climate Change Knowledge Portal (CCKP) in the following website: https://climateknowledgeportal.worldbank.org/. The CCKP serves as a leading hub of global climate data, information, and tools to support decision making and improve the understanding of future climates and projected risks. The CCKP was designed to provide open access to climate data and related information in order to help users better understand future climate scenarios and characterize potential risks and impacts with data aggregated at national, sub-national, and watershed levels (World Bank Group, 2021).

Climate projection data in that website is a modeled data, derived from the Coupled Model Inter-comparison Projects. CMIP data is the foundational data used in the IPCC Assessment Reports; CMIP6 supports the IPPC's Sixth Assessment Report, which it initial release was in August 2021. Projection data in that website is presented at a spatial resolution of 1° × 1° (100 km × 100 km). In CMIP6, future climate scenarios are presented through five SSPs: SSP1-1.9, SSP1-2.6, SSP2-4.5, SSP3-7.0, and SSP5-8.5, which present different societal development pathways. The total radiative forcing level by 2100 (the cumulative measure of GHG emissions from all sources) is presented at the end of each pathway (i.e., – 1.9, – 2.6, – 4.5, 7.0, 8.5) (World Bank Group, 2021).

The description of the two selected future scenarios, namely SSP1-2.6 and SSP5-8.5 is as followed:

SSP1-2.6: supports increasing sustainability with global emissions cut severely, but reach net-zero after 2050.

SSP5-8.5: presents a future based on an intensified exploitation of fossil fuel resources where global markets are increasingly integrated leading to innovations and technological progress (World Bank Group, 2021).

We downloaded the projected monthly data composed of maximum and minimum temperatures as averages of Egypt and sub-region averaged over two time-intervals, namely 2060–2079 and 2080–2099 for SSP1-2.6 and SSP5-8.5 future climate scenarios developed from a multi-model ensemble. From this data, we selected the monthly data of Demiatte governorate, where our analysis was carried out.

Calculation of the Required Irrigation Water for Paddy Rice Under Future Climate

To calculate water requirement of the three studied paddy rice cultivars in Demiatte governorate under the studied future climate scenarios, the monthly evapotranspiration (ETo) values under future climate should be calculated. The downloaded data of each future scenario only contains monthly averages of maximum and minimum temperatures, which were not enough to calculate ETo using Penman–Monteith equation (Allen et al., 1998). However, they are enough to calculate ETo using Hargreaves-Samani equation (Hargreaves & Samani, 1982, 1985). Both ETo equations are included in the BISm model (Snyder et al., 2004), thus, we used it to calculate monthly ETo values of Hargreaves-Samani equation.

The BISm model calculates ETo using the Penman–Monteith equation as presented in Allen et al. (1998). The model also calculates ETo using Hargreaves-Samani equation (Hargreaves & Samani, 1985) if only temperature data are input. Furthermore, the station latitude and elevation must be input for the calculation of ETo. After calculating monthly means of ETo, the model uses a cubic spline curve fitting subroutine to estimate daily ETo rates for the entire year.

To calculate ETo values similar to the values obtained by Penman–Monteith equation using only maximum and minimum temperatures, we used the equation developed by Ouda (2020) for the second agro-climatic zone of Egypt, where Demiatte governorate is located. The equation is stated as followed:

$$\mathrm{ETo[PM\text{ - }HS]} = 0.98 + 1.16 * \mathrm{ETo[HS]} \tag{5.1}$$

where: ETo[PM-HS] is the predicted ETo values similar to the values obtained by Penman-Montieth equation using ETo values resulted from Heregreves-Saminai equation, ETo[HS] is the ETo values obtained from Heregreves-Saminai equation, 0.98 and 1.16 are calibration coefficients represent the intercept (a) and the calibration slope (b) of the best fit regression line, respectively and they considered as regional calibration coefficients for Demiatte governorate.

The downloaded maximum and minimum temperatures of the two selected future climate scenarios in the two time-interval was applied in the BISm model to calculate the ETo values using Heregreves-Saminai equation (ETo[HS]) in the future. These values were applied in Eq. (5.1) to calculated the projected monthly values of ETo, similar to the values obtained by Penman-Montieth equation. These projected ETo values was used to calculate water requirements for the three studied paddy rice cultivars.

To test the strength of the relationship between the measures and projected water requirements, regression analysis between the measured values of water requirements as dependent variable and the projected values as independent variable was done and a prediction equation was developed, as well as the coefficient of determination (R^2) was calculated as followed:

$$R^2 = 1 - \frac{\sum_{i=1}^{n}(O_i - S_i)^2}{\sum_{i=1}^{n}\left(O_i - \overline{O}\right)^2} \quad (5.2)$$

where: O_i and S_i represent the observed and predicted values, respectively, n represents the number of observed and predicted values used in comparison and $\overline{O}$ is the observed average. The R^2 value is calculated from the total sum of squares, more precisely, it is the sum of the squared deviations of the original data from the mean. R^2 takes on values between 0 and 1.0, a value of 1.0 indicates a perfect match, and value of 0 indicates no agreement at all. R^2 also shows how many points fall on the regression line (Jamieson et al., 1998).

Projected Values of the Required Irrigation Water for the Three Studied Paddy Rice Cultivars

Table 5.2 presents the increase in the applied irrigation water for the three paddy rice cultivars under SSP1-2.6 in 2060–2079, compared to the measured applied irrigation water average over the two growing seasons. The results indicated that the required irrigation water will increase by an average of 23 and 20% over the three paddy rice cultivars under irrigation every 6 and 12 days, respectively compared to the measured values.

Figure 5.1a revealed that, under SSP1-2.6 in 2060–2079, a significant relationship ($P < 0.05$) between measured and projected water requirements for the three paddy rice cultivars irrigated every 6 days. Whereas, an insignificant relationship ($P > 0.05$) between measured and projected water requirements for the three paddy rice cultivars irrigated every 12 days was found (Fig. 5.1b).

Similarly, Table 5.3 showed that the average percentage of increase in the required irrigation water for the three paddy rice cultivars under SSP1-2.6 in 2080–2099 will be 23 and 20% under irrigation every 6 and 12 days, respectively compared to the measured values.

Table 5.2 Measured versus projected water requirements (WR, m^3/ha) of the three paddy rice cultivars under SSP1-2.6 future scenario in 2060–2079 and the percentage of increase in the water requirements (PI, %)

	Irrigation every 6 days			Irrigation every 12 days		
	WR	Projected WR	PI (%)	WR	Projected WR	PI (%)
Giza177	13,166	16,260	23	11,261	13,700	22
Giza176	14,530	17,864	23	12,271	14,446	18
Sakha101	14,892	18,237	22	12,595	15,192	21
Average	14,196	17,454	23	12,042	14,446	20

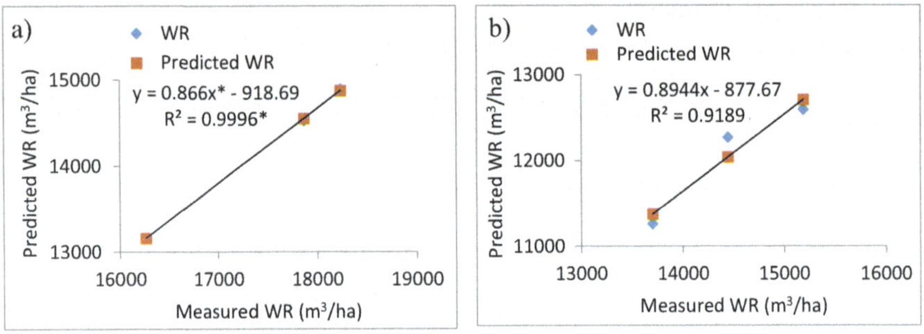

Fig. 5.1 Line fit between measured and projected water requirements (WR) for the three paddy rice cultivars **a** irrigated every 6 days and **b** irrigated every 12 days under SSP1-2.6 in 2060–2079

Table 5.3 Measured versus projected water requirements (WR, m^3/ha) of the three paddy rice cultivars under SSP1-2.6 future scenario in 2080–2099 and the percentage of increase in the water requirements (PI, %)

	Irrigation every 6 days			Irrigation every 12 days		
	WR	Projected WR	PI (%)	WR	Projected WR	PI (%)
Giza177	13,166	16,281	24	11,261	13,682	22
Giza176	14,530	17,889	23	12,271	14,428	18
Sakha101	14,892	18,262	23	12,595	15,174	20
Average	14,196	17,477	23	12,042	14,428	20

A significant relationship ($P < 0.05$) between measured and projected water requirements for rice irrigated every 6 days was found in Fig. 5.2a. Furthermore, an insignificant relationship ($P > 0.05$) between measured and projected water requirements for rice irrigated every 12 days was found in Fig. 5.2b.

With respect to the effect of SSP5-8.5 on the water requirements of paddy rice in 2060–2079, the results in Table 5.4 showed higher increase in it, compared to the values found under SSP1-2.6. The average percentage of increase in the required

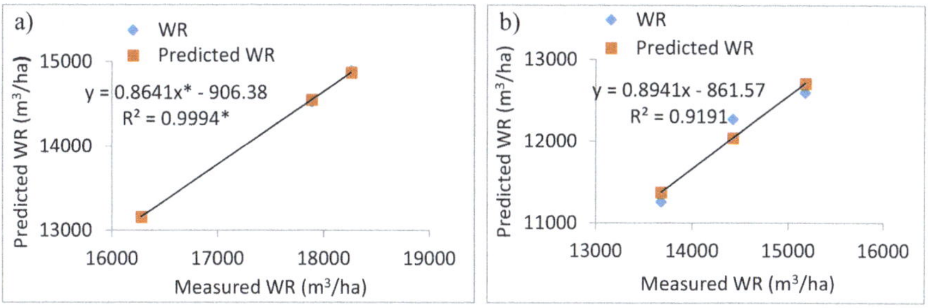

Fig. 5.2 Line fit between measured and projected water requirements (WR) for the three paddy rice cultivars **a** irrigated every 6 days and **b** irrigated every 12 days under SSP1-2.6 in 2080–2099

irrigation water for the three paddy rice cultivars, compared to the measured values is expected to be 29 and 25% under irrigation every 6 and 12 days, respectively.

Under SSP5-8.5 in 2060–2079, Fig. 5.3a revealed a significant relationship ($P < 0.05$) between measured and projected water requirements for the three paddy rice cultivars irrigated every 6 days. Whereas, an insignificant relationship ($P > 0.05$) between measured and projected water requirements for the three paddy rice cultivars irrigated every 12 days (Fig. 5.3b).

Table 5.4 Measured versus projected water requirements (WR, m^3/ha) of the three paddy rice cultivars under SSP5-8.5 future scenario in 2060–2079 and the percentage of increase in the water requirements (PI, %)

	Irrigation every 6 days			Irrigation every 12 days		
	WR	Projected WR	PI (%)	WR	Projected WR	PI (%)
Giza177	13,166	17,057	30	11,261	14,307	27
Giza176	14,530	18,726	29	12,271	15,079	23
Sakha101	14,892	19,112	28	12,595	15,851	26
Average	14,196	18,298	29	12,042	15,079	25

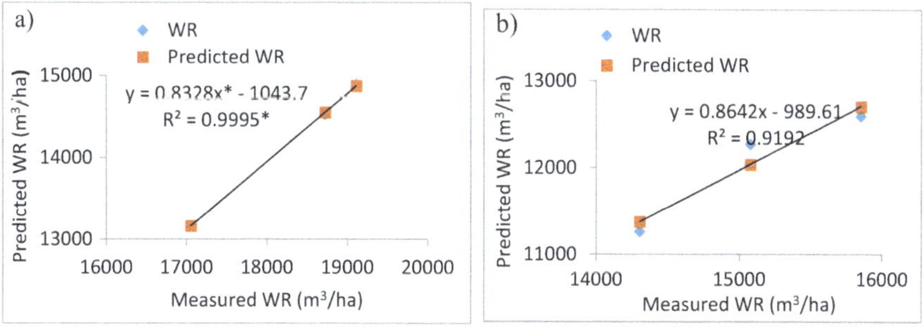

Fig. 5.3 Line fit between measured and projected water requirements (WR) for the three for paddy rice cultivars **a** irrigated every 6 days and **b** irrigated every 12 days under SSP5-8.5 in 2060–2079

A higher average of percentage of increase in the required water to irrigate the three studied paddy rice will be found under SSP5-8.5 in 2080–2099, where it will reach 32 and 28% under irrigation every 6 and 12 days, respectively averaged over the three cultivars, compared to the measured values (Table 5.5).

Similar trend was observed under SSP5-8.5 in 2080–2099, where Fig. 5.4a revealed a significant relationship ($P < 0.05$) between measured and projected water requirements for the three paddy rice cultivars irrigated every 6 days. Whereas, an insignificant relationship ($P > 0.05$) between measured and projected water requirements for the three paddy rice cultivars irrigated every 12 days was found (Fig. 5.4b).

Several studies on the effect of climate change on irrigation water requirements for rice expected an increase in it. Ding et al. (2020) stated that irrigation water requirements of rice, as a water-intensive crop, will be influenced by future climate change, as a result of changes in soil water balances, evapotranspiration, crop physiology, and phenology. Cho et al. (2016) reported an increase in irrigation water requirements using reservoir model HOMWRS and two future scenarios RCP4.5 and RCP8.5 from 2011 to 2040. Using FAO-CropWat model, Acharjee et al. (2017) reported a decrease in crop evapotranspiration and an increase in irrigation water requirements of rice under RCP4.5 and RCP8.5 in 2050 and 2080. Similar results

Table 5.5 Measured versus projected water requirements (WR, m^3/ha) of the three paddy rice cultivars under SSP5-8.5 future scenario in 2080–2099 and the percentage of increase in the water requirements (PI, %)

	Irrigation every 6 days			Irrigation every 12 days		
	WR	Projected WR	PI (%)	WR	Projected WR	PI (%)
Giza177	13,166	17,402	32	11,261	14,636	30
Giza176	14,530	19,117	32	12,271	15,416	26
Sakha101	14,892	19,507	31	12,595	16,196	29
Average	14,196	18,675	32	12,042	15,416	28

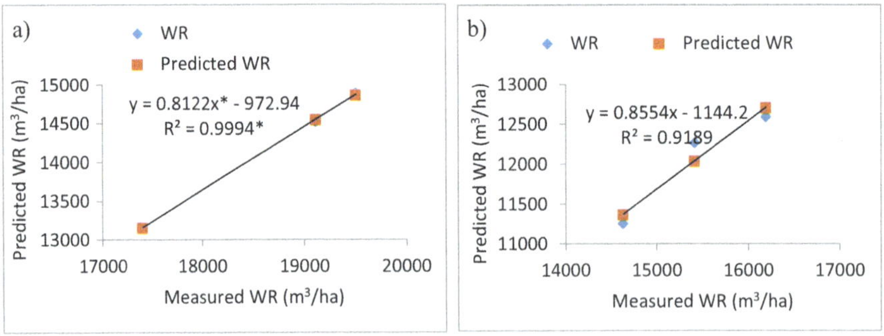

Fig. 5.4 Line fit between measured and projected water requirements (WR) for the three paddy rice cultivars **a** irrigated every 6 days and **b** irrigated every 12 days under SSP5-8.5 in 2080–2099

were obtained by Ding et al. (2020), where they reported an increase in irrigation water requirements of rice using water balance model and three future scenarios, namely RCP2.6, RCP4.5 and RCP8.5 in 2020, 2050, and 2080.

Water Productivity Values of Paddy Rice Under the Studied Future Scenarios

The projected values of water requirements of the three paddy rice cultivars under the studied future scenarios are assumed to be the required amounts in the future to produce the same rice yield of the three cultivars resulted from the field experiment. Thus, these values were used to calculate water productivity. Water productivity is the ratio of yield or marketable product to water used by the crop according to Valipour (2014) as follows:

$$\text{Water productivity(WP)} = \text{Rice yield(kg/ha)/Applied irrigation water}\left(\text{m}^3/\text{ha}\right) \quad (5.3)$$

The values of water productivity of the three paddy rice cultivars under the studied future scenarios are presented in Tables 5.6 and 5.7. The results in Table 5.5 indicated that, in general, water productivity values under irrigation every 6 days will decline under the future climate scenarios, compared to the values obtained under the field experiment conditions (Table 5.1). Similar values of water productivity will be found under SSP1-2.6 in the two time-intervals, i.e. 2060–2079 and 2080–2099, with an average of 0.47 kg/m^3 as a result of low differences in the water requirements under both time-intervals. Whereas, under SSP5-8.5, the situation will be different, where under SSP5-8.5 in 2060–2079, the values of water productivity will be decline, namely 0.45 kg/m^3 and it will be 0.44 kg/m^3 in 2080–2099.

Lower values of water productivity will be obtained under the two studied future scenarios when irrigation was preformed every 12 days. The average values of water productivity over the three cultivars will reach 0.41 kg/m^3 under SSP1-2.6 in the two time-intervals. Whereas, it will be lower under SSP5-8.5, where it will reach 0.39

Table 5.6 The expected water productivity (kg/m^3) under the studied future scenarios for paddy rice cultivars irrigated every 6 days

	SSP1-2.6 (2060–2079)	SSP1-2.6 (2080–2099)	SSP5-8.5 (2060–2079)	SSP5-8.5 (2080–2099)
Giza177	0.46	0.46	0.44	0.43
Giza176	0.48	0.48	0.46	0.45
Sakha101	0.46	0.46	0.44	0.43
Average	0.47	0.47	0.45	0.44

Table 5.7 The expected water productivity (kg/m^3) under the studied future scenarios for paddy rice cultivars irrigated every 12 days

	SSP1-2.6 (2060–2079)	SSP1-2.6 (2080–2099)	SSP5-8.5 (2060–2079)	SSP5-8.5 (2080–2099)
Giza177	0.37	0.37	0.35	0.34
Giza176	0.47	0.47	0.45	0.44
Sakha101	0.39	0.40	0.38	0.37
Average	0.41	0.41	0.39	0.38

and 0.38 kg/m^3 in 2060–2079 and 2080–2099 time-intervals averaged over the three paddy rice cultivars.

Zheng et al. (2020) reported a decrease in the values of water productivity of rice under climate change using crop model ORYZA2000 and RCP2.6, RCP4.5, RCP6.0 and RCP8.5 future scenarios in 2020, 2050, and 2080.

Legume Crops Preceding Paddy Rice

Long-term cropping of cereal crops can degrade soil fertility (Wei et al., 2006), especially when wheat is followed by paddy rice. In paddy rice, nitrogen is one of the most yield-limiting nutrients, due to its low recovery efficiency, which is around 50% in temperate rice systems (Miniotti et al., 2016). The indigenous soil nitrogen supply in paddy rice may decline with intensive rice cultivation. Inclusion of a legume crop in rice-base cropping system can increase soil nitrogen supply through biological nitrogen fixation by associated microbial populations, which can help in reducing soil deterioration (Haefele et al., 2008). Thus, the combined indigenous soil nitrogen and the nitrogen amounts achieved through legume fixation can enrichment soil nitrogen content and increase nitrogen use efficiency in a paddy rice-based cropping system (Thuy et al., 2008). Rahman et al. (2014) planted bush bean, long bean, mung bean, and winged bean preceding paddy rice and reported that rice after winged bean achieved significantly higher nitrogen uptake in both years and produced 13–23% higher grain yield than rice after fallow rotation, which indicated a positive response for rice production without deterioration in soil fertility.

It has been reported that leguminous cover crops, such as hairy vetch can be used as green manure to add nutrients to a rice cropping system and increase soil fertility through biological nitrogen fixation (Prell & Poole, 2006). Lizcano Toledo et al. (2022) stated that the cultivation and incorporation of hairy vetch in the rice cropping system increased grain productivity by 12%, while reducing the applied mineral nitrogen fertilization by 33%. They also reported that the combination of hairy vetch with the incorporation of crop residues, increased soil nitrogen and phosphorous contents, which represented a readily available source for plant uptake over the whole rice cropping season under anaerobic conditions.

Similar benefits were obtained from Persian clover preceding paddy rice. Weinert et al. (2023) reported an increase in the availability of mineral and microbial nitrogen in the soil by 35 and 7%, respectively after the cultivation of Persian clover. The increase in the yield of rice using Persian clover in winter was 8% higher compared to rice yield after fallow.

In Egypt, Sheha et al. (2014) reported an increase in paddy rice yield by 14%, when faba bean preceded it, compare to wheat preceding rice. Whereas, cultivation of fahl clover preceding paddy rice increase rice yield by 3%, compare to wheat preceding rice.

Upland Rice Cultivation in Egypt

Egypt developed a new strategy for rice production to face water scarcity. In addition to maximize water productivity of rice by releasing new short season rice cultivars, with high yielding ability and quality, Egypt also encourage the increasing of the area cultivated with upland rice cultivars to decrease the amount of assigned water to irrigating rice on national level (Abd El-Naby & El-Ghandor, 2022). Due to the prevailing conditions of water shortage in Egypt, there is a remarkable success in breeding new upland rice cultivars requiring less irrigation water amounts than paddy rice cultivars (Gaballah et al., 2020). Furthermore, there are massive efforts for expanding the cultivation of the upland rice cultivars released in Egypt to replace the paddy rice cultivars, in order to conserve water resources (Sheha et al., 2021).

Effect of Deficit Irrigation on the Yield of Upland Rice Under Climate Change

The Field Experiments

The effect of application of deficit irrigation to upland rice under climate change scenarios was assessed through a two-year field experiment implemented at El-Behira Governorate (Lat. 30.85° N, Long. 30.34° E, 23 m a.s.l.); Egypt. The objective was to compare between the effect of the applied irrigation amount to upland rice by the farmer every 6 days (FP) and our applied irrigation treatments every 6 days using 100% ETc (RI) on the final rice yield under current and future climate. Irrigation of rice every 6 days is the common practice in the surrounding area due to water availability in the irrigation canals. Furthermore, the effect of two deficit irrigation treatments (irrigation every 9 days (DI1) and irrigation every 12 days (DI2)) using 100% ETc on the yield of upland rice was also determined. This assessment was done for upland rice grown under current climate (2019 and 2020 growing seasons)

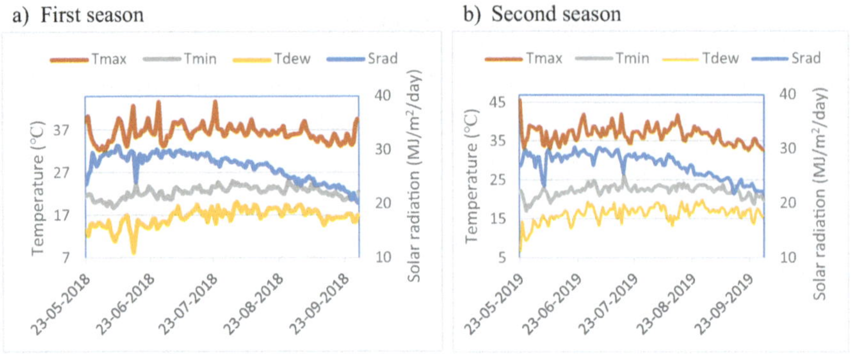

Fig. 5.5 The daily values of solar radiation, and maximum, minimum and dewpoint temperatures in the first growing season (**a**) and in the second growing season (**b**)

and under two and future climate scenarios, namely SSP1-2.6 and SSP5-8.5 in two time-intervals, i.e. 2060–2079 and 2080–2099.

The daily meteorological data of the experimental site in 2018 and 2019 were obtained from the following website: https://power.larc.nasa.gov/data-access-viewer/. The values of the daily solar radiation (S_{rad}, MJ/m^2/day), maximum temperature (T_{max}, °C), minimum temperature (T_{min}, °C) and dewpoint temperature (T_{dew}, °C) are presented in Fig. 5.5.

The soil of the experimental site was clayey, with 60.56% clay, 24.50% silt and 15.82% sand averaged over 60 cm of soil depth. The field capacity was 48.13%, available water was 26.11% and wilting point was 22.02% averaged over 60 cm of soil depth.

Ranonamized complete blocks design with three replicates was used, where the three irrigation water intervals (RI, DI1 and DI2) were randomly assigned in each block. The applied irrigation by the farmer was measured separately from the experimental design, as well as rice yield obtained by the farmer and were used for comparison purposes only.

Land ploughing was done twice, then the land was divided to plots, its area was 42.0 m^2 (6.0 m × 7.0 m). The rice seeds were planted in rows in basins, where basins were isolated from the each other to prevent horizontal water movement. To avoid the lateral movement of water and for more water control, 2 m wide ditches separated the plots.

Potassium sulfate (48% K_2O) was applied in the rate of 60 kg/ha and calcium super phosphate (15.5% P_2O_5) was applied at a rate of 480 kg/ha during land preparation. The upland rice cultivar Giza183 was sown in dry soil after soaking in water for 24 h, in 22nd and 23rd of May in the first and second growing season, respectively. Nitrogen fertilizer was added with 192.0 kg N/ha in three doses and 60 kg/ha of potassium sulfate (48% K_2O) was applied 60 days after planting.

Surface irrigation was used to deliver water to the experiment. The applied irrigation water amounts estimating crop evapotranspiration or irrigation amounts under standard conditions was calculated using the following equation (Allen et al., 1998):

$$\mathrm{ETc} = \mathrm{ETo} * K_c \tag{5.4}$$

where: ETc = crop evapotranspiration (mm/day), ETo = reference crop evapotranspiration (mm/day), and K_c = crop coefficient.

ETo was calculated using the Penman–Monteith equation (Allen et al., 1998). The equation uses averages of the standard meteorological data of daily measurements of solar radiation (sunshine hours), air temperature (maximum and minimum), relative humidity, and wind speed. The values of K_c of rice were obtained from BISm model (Snyder et al., 2004), and used to calculate the depth of the required irrigation water according to Allen et al. (1998).

The amount of required irrigation water was calculated according to the following equation (Vermeiren & Jopling, 1984):

$$\mathrm{AIW} = \frac{\mathrm{ETo} \times I}{\mathrm{Ea}(1 - \mathrm{LR})} \tag{5.5}$$

where: AIW = applied irrigation water depth (mm/day), Ea = irrigation application efficiency equals 60% under surface irrigation system. LR = leaching requirements (was not considered).

Harvest was done 29th and 30th of September in the first and second growing season, respectively. The yield of each plot was harvested, weighted and added together to measure grain yield (ton/ha).

The ETo Values and the Applied Irrigation Water to Upland Rice

Figure 5.6a presents a comparison between the values of ETo in both growing seasons of upland rice. The figure revealed that the ETo values in May and August were higher in the second growing season, compared to the first growing season. Figure 5.5b showed that the values of maximum and minimum temperatures in May and August were higher in the second growing season, compared to its values in the first growing season, which could have reflected on the values of ETo and increase it.

Figure 5.6b also presents a comparison between values of the amounts of the corresponded applied irrigation water in both growing seasons of upland rice. The figure showed that the applied irrigation water was slightly higher in the second growing season than the value applied in the first growing season.

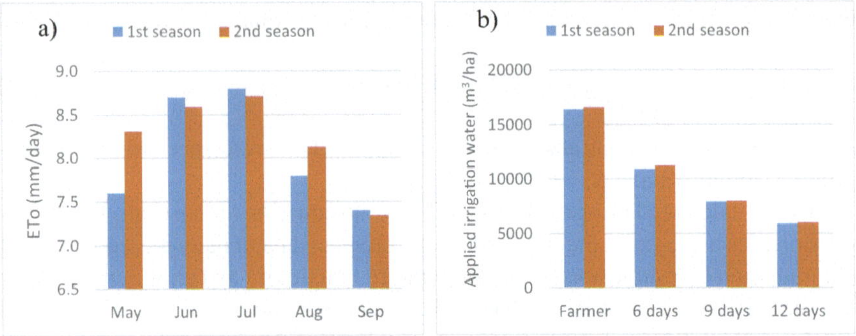

Fig. 5.6 The values of ETo (**a**) and the corresponding applied irrigation amounts (**b**) in both growing seasons

The Yield of Rice and Its Water Productivity

Table 5.8 presents comparison between farmer practice in irrigating rice and our irrigation treatments. It is clear from Table 5.8 that the highest applied irrigation amounts were done by the farmer and resulted in lower yield, compared to irrigation every 6 and 9 days using 100% ETc. The yield obtained by the farmer was lower by 29 and 7% than the values obtained under irrigation every 6 and 9 days, respectively averaged over the two growing seasons. Whereas, rice yield obtained under irrigation every 12 days was lower by 16% than the yield obtained by the farmer and it was lower by 36% than the yield obtained under irrigation every 6 days. Furthermore, irrigation every 9 and 12 days reduced rice yield by 17 and 35%, respectively compare to irrigation every 6 days.

Sheha et al. (2021) reported a reduction in the yield of upland rice by 15 and 36% under irrigation every 9 and 12 days, compared to the yield obtained under irrigation every 6 days. Alou et al. (2018) attributed the loss in the yield of upland rice by being sensitive to water stress during tillering, floret initiation and subsequent

Table 5.8 The yield of upland rice and water productivity as affected by irrigation treatments in both growing seasons

Irrigation	First growing season		Second growing season	
	Yield (ton/ha)	Water productivity (kg/m^3)	Yield (ton/ha)	Water productivity (kg/m^3)
FP	7.54	0.46	7.45	0.45
RI	9.66	0.89	9.59	0.85
DI1	8.02	1.01	7.96	1.00
DI2	6.28	1.07	6.23	1.04

FP = farmer practice, RI = irrigation every 6 days, DI1 = irrigation every 9 days, DI2 = irrigation every 12 days

spikelet sterility, as well as during grain filling stage. Drought before and after heading significantly impacts brown and milled rice rates by lowering its quality (Mohamed et al., 2019).

Consequently, water productivity was the lowest under farmer practice due to the excessive applied irrigation amount and low yield, where it was 0.46 and 0.45 kg/m^3 in the first and second growing season, respectively (Table 5.7). It has been reported that application excessive irrigation led to soil erosion and contamination of surface and subsurface water with nitrates, and phosphorous due to runoffs (Bilgili, 2013), and that could be the reason for lower rice yield under farmer practice. Furthermore, the highest values of water productivity were obtained under irrigation every 12 days, namely 1.07 and 1.04 kg/m^3 in the first and second season, respectively (Table 5.8).

Climate-Suitability of Upland Rice Cultivation in the Future

Temperature is the factor that limits the geographical areas where different crops can be grown, as well as being the main weather element that control the rate of development, growth and crop yields (Rehman et al., 2015). Agricultural crops have basic temperature requirements to complete a particular phenological phase, as well as its entire life cycle (Luo, 2011). Rice plant requires specific optimum temperatures at different phenological stages during its growing season (Kim et al., 2011). However, temperature above a threshold upper level may reduce phenological phases and resulted in lesser productivity and quality (Ahmad et al., 2015). The expected rise in the temperature under climate change could cause stressful conditions during the growing season of upland rice in a certain region. This could lead to lower rice productivity to an extent, where it will be not economical to cultivate it in this region.

To test this assumption, we used the downloaded projected monthly maximum and minimum temperatures as averages in 2060–2079 and 2080–2099 for SSP1-2.6 and SSP5-8.5 future climate scenarios developed from a multi-model ensemble for El-Behira governorate, where our experiment was carried out. Maximum temperature under the two studied future climate scenarios in the two time-interval was compared to the cutoff temperature of rice, namely 35 °C (Yoshida, 1981) to determine the days where maximum temperature will exceed the cutoff temperature.

Figure 5.7 showed that the frequency of temperature exceeding 35 °C will be lower under SSP1-2.6 in 2060–2079 (Fig. 5.7a), than the frequency that will occur in 2080–2099 (Fig. 5.7b) under the same future scenario, which will occur during mid-season of upland rice. This result implied that rice productivity will be negatively affected in these two time-intervals.

The frequency of temperature exceeding 35°C under SSP5-8.5 will be higher than its incidence under SSP1 2.6, where most of the growing season of upland rice will experience higher temperature than the cutoff temperature (Fig. 5.8). Figure 5.8a showed that, during 2060–2079, the rise in temperature will be mostly between 35 and 40 °C, except for one time it will exceed 40 °C. Whereas during 2080–2099, Fig. 5.8b showed that the temperature will exceed 40 °C most of the upland rice

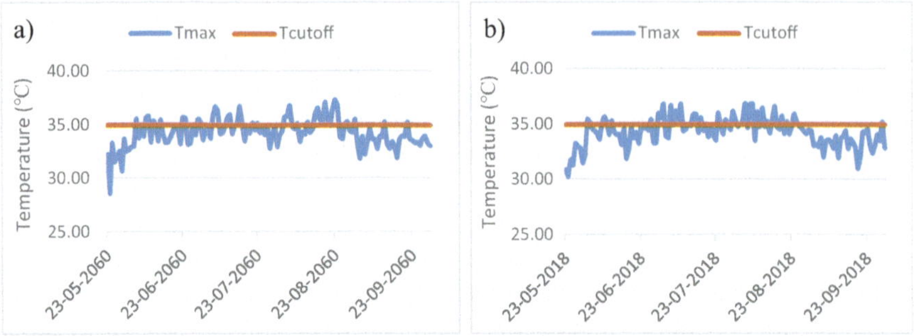

Fig. 5.7 Comparison between projected maximum temperature and cutoff temperature for upland rice in El-Behira governorate under SSP1-2.6 in **a** 2060–2079 and **b** 2080–2099

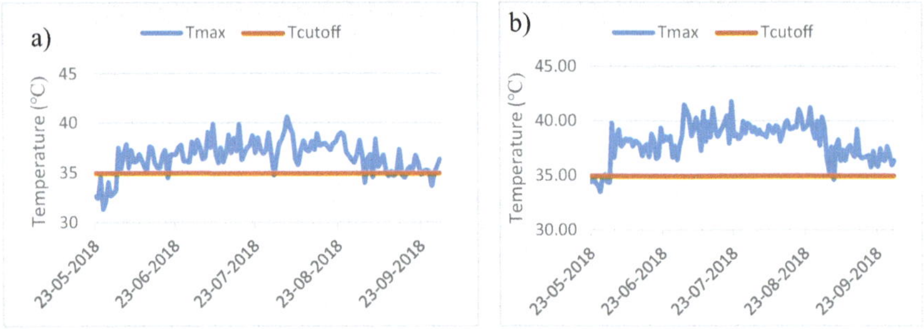

Fig. 5.8 Comparison between projected maximum temperature and cutoff temperature for rice in El-Behira governorate under SSP5-8.5 in **a** 2060–2079 and **b** 2080–2099

growing season, which implied higher yield losses than the values obtained under SSP1-2.6.

Similar results were obtained by Mahmoud et al. (2016) in El-Behira governorate using A1B climate change scenario developed by ECHAM5 climate model in 2040, where they reported an increase in the number of days temperature exceeding 35 ºC during rice growing season.

Calculation of Water Requirements of Upland Rice Under Future Climate

Hong et al. (2016) expected a variation in irrigation water requirements for rice due to changes in evapotranspiration values under different future climate scenarios, and that depend on cultivar, soil, and weather characteristics. Irrigation water requirement of upland rice under the studied future climate scenarios, were calculated from monthly ETo values under future climate using Eq. (5.1). We applied the obtained

ETo values in Eq. (5.4) and the required irrigation water for upland rice was calculated under the studied future climate scenarios. R^2 between the values of measured water requirements and the projected values was calculated. Additionally, to test the goodness of fit between the measured and the predicted values, Willmott index of agreements (*d-stat*) and root mean square error (RMSE) were calculated as followed:

Willmott Index of Agreement (**d-stat**)

It is the standardized measure of the degree of model prediction error and varies between 0 and 1.0. A value of 1.0 indicates a perfect match, and value of 0 indicates no agreement at all (Willmott, 1981).

$$d\text{-}stat = 1 - \frac{\sum_{i=1}^{n}(O_i - S_i)^2}{\sum_{i=1}^{n}\left[\left(\left|(S_i - \overline{O})\right| + \left|(O_i - \overline{O})\right|\right)^2\right]} \quad (5.6)$$

Root Means Square Error (RMSE)

It is the standard deviation of the residuals (prediction errors). Residuals are a measure of how far from the regression line data points are. It is a frequently used to measure the difference between values predicted by a model and those actually observed from an experiment that is being modeled (Jamieson et al., 1998).

$$\text{RMSE} = \sqrt{\left(\frac{\sum_{i=1}^{n}(S_i - O_i)^2}{n}\right)} \quad (5.7)$$

Projected Water Requirements Under the Studied Climate Future Scenarios

Table 5.9 showed that under SSP1-2.6 in 2060–2079, the required irrigation water will increase by a value between 15 and 19%, with the highest value will be obtained under irrigation every 6 days and the lowest value will be under irrigation every 12 days. Under farmer practice, the percentage of increase will be lower than the percentage obtained with irrigation every 6 days and it will be higher than the obtained value under irrigation every 9 days.

Table 5.9 Comparison between the measured irrigation water requirements (WR) of upland rice and the projected values obtained under SSP1-2.6 future scenario and the percentage of increase (PI)

	SSP1-2.6 (2060–2079)			SSP1-2.6 (2080–2099)		
	Measured WR (m^3/ha)	Projected WR (m^3/ha)	PI (%)	Measured WR (m^3/ha)	Projected WR (m^3/ha)	PI (%)
FP	16,390	19,107	17	16,390	19,508	19
RI	10,870	12,927	19	10,870	13,077	20
DI1	7911	9138	16	7911	9311	18
DI2	5895	6806	15	5895	6938	18
R^2	0.9992			0.9997		
d-stat	0.9547			0.9445		
RMSE (m^3/ha)	2156			2425		

FP = farmer practice, RI = irrigation every 6 days, DI1 = irrigation every 9 days, DI2 = irrigation every 12 days

Similar trend will be obtained under SSP1-2.6 in 2080–2099 time-interval with higher values of percentage of increase in water requirements of rice, where it will be between 18 and 20%, with the highest value will be obtained under irrigation every 6 days.

The table also showed that both the values of R^2 and *d-stat* were high in both time-intervals. Furthermore, the values of RMSE were also high in both time-intervals, which imply the high deviation between the measure water requirements and the projected values.

A significant positive relationship ($P < 0.01$) between the measured applied irrigation water (WR) to upland rice and its corresponded projected values was found for each of the studied time-interval (Fig. 5.9). The developed prediction equations could be used to project the required irrigation for upland rice at El-Behira governorate under SSP1-2.6 in 2060–2079 (Fig. 5.9a) and in 2080–2099 (Fig. 5.9b).

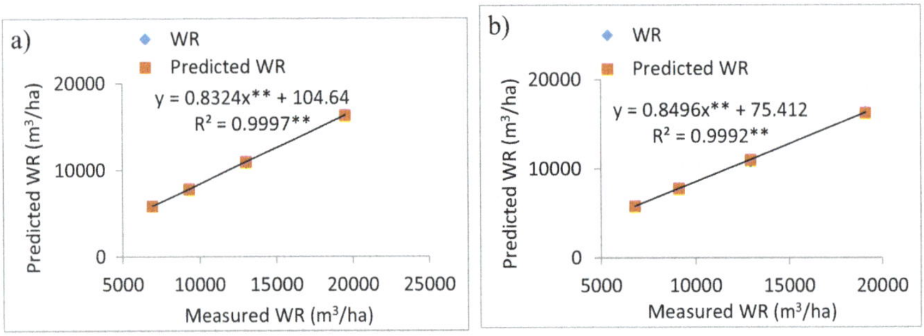

Fig. 5.9 Line fit between measured water requirements (WR) and the projected values under SSP1-2.6 in 2060–2079 (**a**) and 2080–2099 (**b**)

Table 5.10 Comparison between the measured irrigation water requirements (WR) of upland rice and the projected values obtained under SSP5-8.5 future scenario and the percentage of increase (PI)

	SSP5-8.5 (2060–2079)			SSP5-8.5 (2080–2099)		
	Measured WR	Projected WR	PI (%)	Measured WR	Projected WR	PI (%)
FP	16,390	19,944	22	16,390	20,425	25
RI	10,870	13,417	23	10,870	13,753	27
DI1	7911	9564	21	7911	9801	24
DI2	5895	7122	21	5895	7287	24
R^2	0.9996			0.9995		
d-stat	0.9291			0.9124		
RMSE	2790.2			3167.7		

FP = farmer practice, RI = irrigation every 6 days, DI1 = irrigation every 9 days, DI2 = irrigation every 12 days

Table 5.10 showed a similar trend in the percentage of increase between measured and projected water requirements under SSP5-8.5 in both time-intervals to the one obtained under SSP1-2.6 in both time intervals (Table 5.9), where the highest value will be obtained under irrigation every 6 days and the lowest value will be under irrigation every 12 days. Under farmer practice, the percentage of increase will be lower than the percentage obtained with irrigation every 6 days and it will be higher that the obtained under irrigation every 9 days. The required irrigation water will increase between 21–23% in 2060–2079 time-interval and between 24–27% in 2080–2099 time-interval.

The table also showed that both the values of R^2 and *d-stat* were high in both time-intervals, as well as the values of RMSE in both time-intervals, which imply the high deviation between the measure water requirements and the projected values.

A significant positive relationship ($P < 0.01$) between the measured water requirements (WR) to upland rice and its corresponded projected values was found for each of the studied time-interval (Fig. 5.10). The developed prediction could be used to project the required irrigation for upland rice in El-Behira governorate under SSP5-8.5 in 2060–2079 (Fig. 5.10a) and in 2080–2099 (Fig. 5.10b).

Similar results were obtained by Mahmoud et al. (2016) in El-Behira governorate using A1B climate change scenario in 2040 developed by ECHAM5 climate model, where they reported an increase in the required water requirement for rice by 11%.

Water Productivity of Upland Rice Under Climate Change

Losses in the values of water productivity of upland rice will occur under the studied future climate scenarios (Table 5.11), compared to the values obtained under the field experiment (Table 5.8). The farmer practice will attain the lowest value of

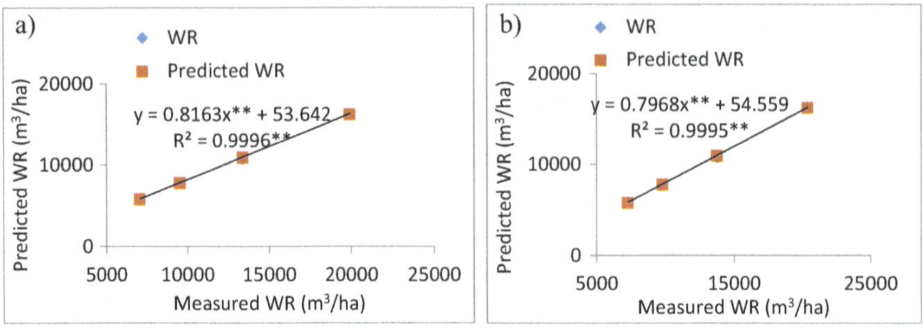

Fig. 5.10 Line fit between measured water requirements (WR) and the projected values under SSP5-8.5 in 2060–2079 (**a**) and 2080–2099 (**b**)

Table 5.11 Water productivity (kg/m^3) of upland rice under the studied future climate scenarios averaged over the two growing seasons

	SSP1-2.6 (2060–2079)	SSP1-2.6 (2080–2099)	SSP5-8.5 (2060–2079)	SSP5-8.5 (2080–2099)
FP	0.38	0.39	0.38	0.37
RI	0.74	0.74	0.72	0.70
DI1	0.86	0.87	0.84	0.82
DI2	0.90	0.92	0.88	0.86

FP = farmer practice, RI = irrigation every 6 days, DI1 = irrigation every 9 days, DI2 = irrigation every 12 days

water productivity under the two studied future scenarios in both time-intervals. This could be due to the high applied irrigation water aby the farmer and its low corresponded rice yield. The results in the table also showed that the highest values of water productivity will be obtained under irrigation every 12 days under the two studied future scenarios in both time-intervals, where lower amount of irrigation water was applied, compared to the required irrigation (irrigation every 6 days) and also lower rice yield was obtained.

Simulation of the Effect of Deficit Irrigation on Upland Rice Yield Under Climate Change Using SALTMED Model

Simulation models are valuable tools to assist in the impact of climate change on different crops (Ewert, 2012). SALTMED (Ragab, 2002, 2015) is one of the most important process-oriented crop simulation models. The model uses integrated crop-soil–water processes and have been proven to be an effective tool for water resources management in agriculture (Ragab, 2015). The model has been widely used for different crops irrigated with different irrigation and fertigation systems implemented in different parts of the world. It has been used in Egypt and Syria for tomato crop (Ragab et al., 2005a, 2005b), for sugarcane in Iran (Golabi et al., 2009), and for chickpea in Portugal (Silva et al., 2013). It has been also used for quinoa in Morocco (Fghire et al., 2015; Hirich et al., 2012), in Denmark (Razzaghi et al., 2011), and in Italy (Pulvento et al., 2013). It has been used for amaranth in Italy (Pulvento et al., 2015), and for legumes in Syria (Arslan et al., 2015). A detailed description of the SALTMED model have been provided by Ragab (2002, 2015) and Ragab et al. (2005a, 2005b).

The SALTMED model was calibrated for the first growing season data of upland rice grown under the field experimental and the data of the second growing season was used to validate the model. The calibration was done by fine tuning of "photosynthesis efficiency factor" to obtained reasonable values of the yield under the studied treatments, whereas the validation was done by applying the calibrated values in the model to predict the yield under the studied treatments in the second season.

The SALTMED model requires to input soil file, crop file, climate file, and nitrogen file. The soil file was prepared using the soil data of the experimental site. A weather file was prepared starting from the sowing date of upland rice and ended with the harvest date in each growing season.

For the crop file required by the model, one crop file was prepared for each growing season. The input data required by SALTMED was obtained from measured field data (planting date, leaf area index, plant height, and the length of the growth stages). The other input data required by the model was obtained from the literature (root depth, Kc and Kcb) (Allen et al., 1998). Furthermore, we used the function "Fixed length growth stages" where harvest day is predetermined by the user.

The model calculates the reference evapotranspiration (ETo) by several methods among them is the FAO modified Penman–Monteith equation according to Allen et al. (1998), which was selected for this study. Furthermore, in each growing season, one irrigation file was prepared for each irrigation treatment under basin irrigation.

Regarding plant growth process for biomass calculation, SALTMED provides a detailed physiological imbedded factors in the "Crop growth tab" to be used in the calculation, or to be modified by the user.

Calibration and Validation of SALTMED Model

Table 5.12 presents the results of the goodness of fit test for both calibrated and validated values in both growing seasons. The result in the table proved that SALTMED model was able to simulate the growth environment of upland rice under the field experiments with high level of accuracy as it was demonstrated by high value of R^2, and *d-stat*, as well as low value of RMSE in both calibrated and validated growing seasons.

A significant positive relationship ($P < 0.01$) between the measured upland rice yield and its corresponded simulated values by the SALTMED model was found for each of the studied growing season (Fig. 5.11).

Table 5.12 Measured versus predicted upland rice yield (ton/ha) in both growing seasons, and percentage of yield reduction (PR%)

	First growing season (calibrated)			Second growing season (validated)		
	Measured yield	Predicted yield	PR (%)	Measured yield	Predicted yield	PR (%)
FP	7.54	7.35	3	7.45	7.16	4
RI	9.66	9.25	4	9.59	9.08	5
DI1	8.02	7.66	4	7.96	7.47	6
DI2	6.28	6.09	3	6.23	5.98	4
R^2	0.9981			0.9973		
d-stat	0.9855			0.9707		
RMSE (ton/ha)	0.3266			0.4649		

FP = farmer practice, RI = irrigation every 6 days, DI1 = irrigation every 9 days, DI2 = irrigation every 12 days

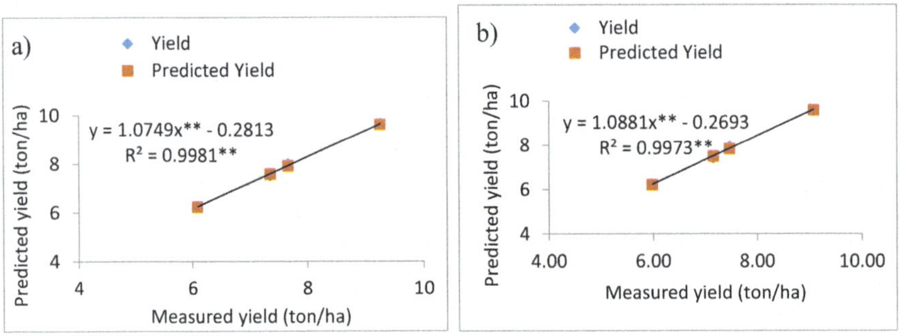

Fig. 5.11 Line fit between measured upland rice yield in the first growing season (**a**) and the second growing season (**b**)

Calculation of Missing Climate Elements to Calculate ETo Under Future Climate Scenarios

The downloaded data of the future scenarios is composed of monthly averages of maximum and minimum temperature only. The SALTMED model uses daily data to run the model, therefore the monthly data was change to daily data using MS-Excel "Random number generation" function. The model also requires to input relative humidity, sunshine hours and wind speed to calculate ETo and dry matter accumulation.

To calculate relative humidity, we used the equation presented by Allen et al. (1998). The relative humidity (RH) expresses the degree of saturation of the air as a ratio of the actual vapor pressure (e_a) to the saturation vapor pressure (e_o) as followed:

$$e_a = 0.611 * \exp((17.27 * T_{\min})/(T_{\min} + 237.3)) \tag{5.8}$$

$$e_o = 0.618 * \exp((17.27 * T_{\max})/(T_{\max} + 237.3)) \tag{5.9}$$

$$\mathrm{RH} = 100 * (e_a/e_o) \tag{5.10}$$

where: $T_{\min}$ and $T_{\max}$ are minimum and maximum temperature (°C), respectively.

We also calculated sunshine hours (SSH) using the equation presented by Abd El-Wahed and Snyder (2015) as followed:

$$\mathrm{SSH} = 4.352 + 0.232(T_{\mathrm{mean}}) \tag{5.11}$$

where: T_{mean} is mean temperature (°C)

With respect to wind speed, Allen et al. (1998) stated that "the variation in wind speed average over monthly periods is relatively small and fluctuates around average values. The 'average' wind speed estimates may be selected from information available for the regional climate". Therefore, the wind speed data under current climate was used to substitute for its absence in the future.

Projection of Upland Rice Yield Under Future Climate Scenarios Using SALTMED Model

Future Climate Scenario SSP1-2.6 in 2060–2079

Table 5.13 presents the projection of the yield of upland rice in both growing seasons using SSP1-2.6 in 2060–2079. The results in the table showed that rice yield resulted from farmer practice was lower than the yield obtained under required irrigation, where the percentage of reductions between measured and projected yield were 9% versus 8%, respectively under SSP1-2.6 averaged over the two growing seasons. Irrigation every 9 days increase yield losses to reach 10%, whereas irrigation every 12 days resulted in higher value of yield losses, namely 11% averaged over the two growing seasons.

The goodness of fit analysis revealed high values of R^2 namely 0.9984 and 0.9986 in the first and second growing season, respectively. The analysis also showed high values of *d-stat*, namely 0.9169 and 0.9158 in the first and second season, respectively. RMSE values between measured and projected yield were 0.8439 and 0.8323 ton/ha in the first and second growing season, respectively, which implied high differences between the measured and projected yields values under the SSP1-2.6 in 2060–2079 (Table 5.13).

Table 5.13 Measured versus projected upland rice yield (ton/ha) in the two growing seasons, and percentage of yield reduction (PR%) using SSP1-2.6 future scenario in 2060–2079

	First growing season			Second growing season		
	Measured	Projected	PR (%)	Measured	Projected	PR (%)
FP	7.54	6.88	9	7.45	6.78	9
RI	9.66	8.92	8	9.59	8.82	8
DI1	8.02	7.22	10	7.96	7.16	10
DI2	6.28	5.56	11	6.23	5.60	10
R^2	0.9984			0.9986		
d-stat	0.9169			0.9158		
RMSE (ton/ha)	0.8439			0.8323		

FP = farmer practice, RI = irrigation every 6 days, DI1 = irrigation every 9 days, DI2 = irrigation every 12 days

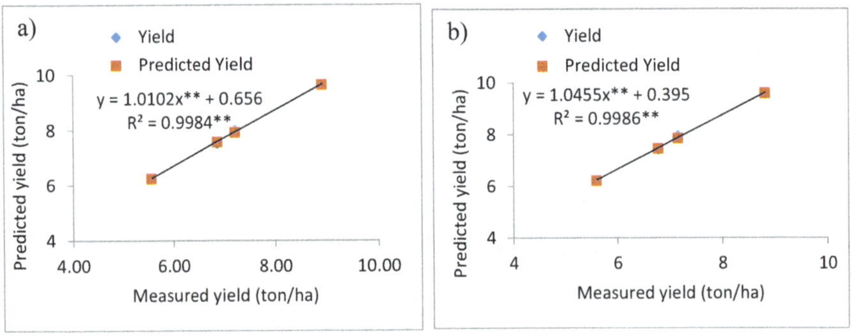

Fig. 5.12 Line fit between the measured upland rice yield and its corresponded projected values by SALTMED model in the first growing season (**a**) and in the second growing season (**b**) under SSP1-2.6 in 2060–2079

A significant positive relationship (*P* < *0.001*) was found between the measured upland rice yield and its corresponded projected values by the SALTMED model in both growing season (Fig. 5.12).

Future Climate Scenario SSP1-2.6 in 2080–2099

Because SSP1-2.6 describes a future with increasing sustainability and highly cut in the global emissions reaching net-zero after 2050, the reductions in the yield of upland rice in 2080–2099 (Table 5.14) were close to the values obtained in 2060–2079 (Table 5.13). The results in Table 5.14 also showed that the highest reduction in the yield of upland rice will occur under irrigation every 12 days, namely 12%, and the lowest yield reduction will occur under the application of required irrigation, namely 7% averaged over the two growing seasons.

The goodness of fit analysis showed high values of R^2, namely 0.9968 and 0.9958 in the first and second growing season, respectively. It also showed high values of *d-stat*, namely 0.9182 and 0.9190 in the first and second growing season, respectively. Furthermore, the values of RMSE between the measured and projected yield were 0.8491 and 0.8333 ton/ha in the first and second growing season, respectively (Table 5.14).

The regression analysis between the measured and projected yield of upland rice by the SALTMED model showed a significant positive relationship ($P < 0.001$) in both growing seasons (Fig. 5.13).

Table 5.14 Measured versus projected upland rice yield (ton/ha) in both growing seasons, and percentage of yield reduction (PR%) using SSP1-2.6 future scenario in 2080–2099

	First growing season			Second growing season		
	Measured	Projected	PR (%)	Measured	Projected	PR (%)
FP	7.54	6.86	9	7.45	6.75	9
RI	9.66	9.00	7	9.59	8.93	7
DI1	8.02	7.17	11	7.96	7.11	11
DI2	6.28	5.54	12	6.23	5.57	11
R^2	0.9968			0.9958		
d-stat	0.9182			0.9190		
RMSE (ton/ha)	0.8491			0.8333		

FP = farmer practice, RI = irrigation every 6 days, DI1 = irrigation every 9 days, DI2 = irrigation every 12 days

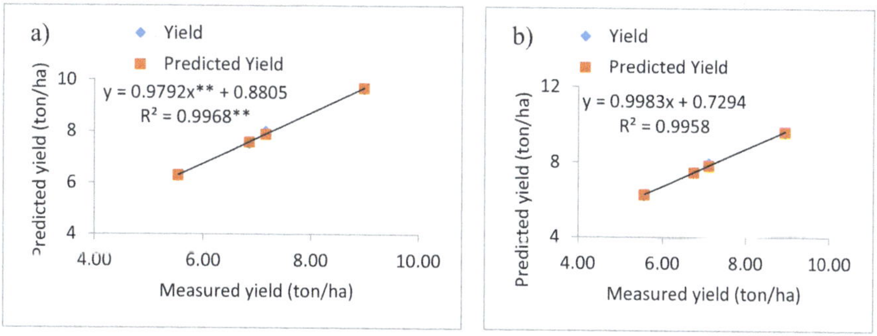

Fig. 5.13 Line fit between the measured upland rice yield and its corresponded projected values by SALTMED model in the first growing season (**a**) and in the second growing season (**b**) under SSP1-2.6 in 2080–2099

Table 5.15 Measured versus projected yield of upland rice (ton/ha) in the two growing seasons, and percentage of yield reduction (PR%) using SSP5-8.5 future scenario in 2060–2079

	First growing season			Second growing season		
	Measured	Projected	PR (%)	Measured	Projected	PR (%)
FP	7.54	6.58	13	7.45	6.58	12
RI	9.66	8.90	8	9.59	8.80	8
DI1	8.02	7.03	12	7.96	7.10	11
DI2	6.28	5.47	13	6.23	5.51	12
R^2	0.9942			0.9956		
d-stat	0.8863			0.9048		
RMSE (ton/ha)	1.0213			0.8924		

FP = farmer practice, RI = irrigation every 6 days, DI1 = irrigation every 9 days, DI2 = irrigation every 12 days

Future Climate Scenario SSP5-8.5 in 2060–2079

Because SSP5-8.5 represents stressful conditions, where it is based on an intensified exploitation of fossil fuel resources, it is expected that more yield losses will occur under its assumption. The simulation results projected higher reduction in the yield of upland rice under all treatments, compared to the reduction occurred under SSP1-2.6. The highest yield reduction will be obtained under farmer practice and irrigation every 12 days, namely 13% averaged over the two growing seasons. Whereas, the lowest yield reduction will be obtained under application of required irrigation, namely 8% in both growing seasons (Table 5.15).

High values of R^2, namely 0.9942 and 0.9956 in the first and second growing season, respectively were found by the goodness of fit test. The analysis also showed high values of *d-stat*, namely 0.8863 and 0.9048 in the first and second growing season, respectively. Moreover, high values of RMSE between measured and projected yield were found, namely 1.0213 and 0.8924 ton/ha in the first and second growing season, respectively (Table 5.15).

A significant positive relationship ($P < 0.001$) was found between the measured upland rice yield and its corresponded projected values by the SALTMED model in both growing seasons (Fig. 5.14).

Future Climate Scenario SSP5-8.5 in 2080–2099

Higher reduction in the yield of upland rice will occur under SSP5-8.5 future climate scenario in 2080–2099 (Table 5.16). The farmer practice and irrigation every 12 days will attain the highest yield losses, namely 14% averaged over both growing seasons.

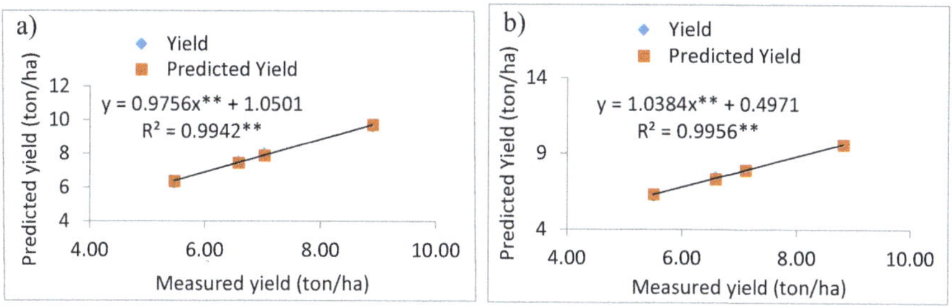

Fig. 5.14 Line fit between the measured upland rice yield and its corresponded projected values by SALTMED model in the first growing season (**a**) and in the second growing season (**b**) under SSP5-8.5 in 2060–2079

Whereas, the lowest yield losses will occur under application of required irrigation, namely 9%.

The goodness of fit analysis revealed high values of R^2 namely 0.9933 and 0.9974 in the first and second growing season, respectively. It also showed high values of *d-stat*, namely 0.8670 and 0.8834 in the first and second growing season, respectively. Additionally, RMSE values between measured and projected yield were 1.1168 and 1.0160 ton/ha in the first and second growing season, respectively (Table 5.16).

A significant positive relationship ($P < 0.001$) was found between the measured upland rice yield and its corresponded projected values by the SALTMED model in both growing seasons (Fig. 5.15).

Mahmoud et al. (2016) indicated that rice yield could decrease by 20% in El-Behira governorate under A1B climate change scenario developed by ECHAM5 climate model in 2040.

The decline in rice yield under future climate scenario will be existed as a result of increasing spikelet sterility (Dubey et al., 2018). The rise in temperature increases

Table 5.16 Measured versus projected yield of upland rice (ton/ha) in the two growing seasons, and percentage of yield reduction (PR%) using SSP5-8.5 future scenario in 2080–2099

	First growing season			Second growing season		
	Measured	Projected	PR (%)	Measured	Projected	PR (%)
FP	7.54	6.48	14	7.45	6.50	13
RI	9.66	8.82	9	9.59	8.74	9
DI1	8.02	6.95	13	7.96	7.04	12
DI2	6.28	5.40	14	6.23	5.44	13
R^2	0.9933			0.9974		
d-stat	0.8670			0.8834		
RMSE (ton/ha)	1.1168			1.0160		

FP = farmer practice, RI = irrigation every 6 days, DI1 = irrigation every 9 days, DI2 = irrigation every 12 days

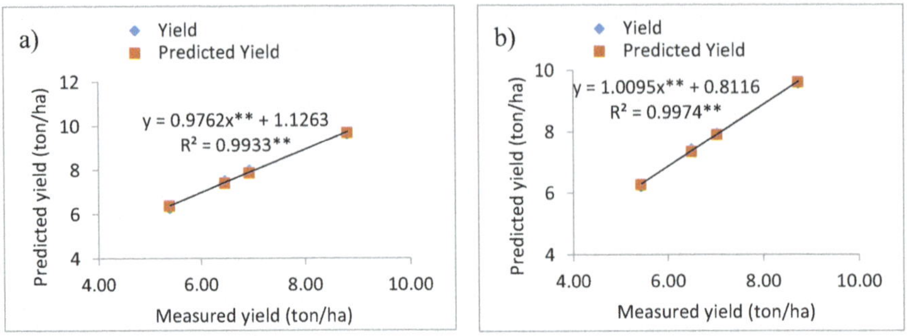

Fig. 5.15 Line fit between the measured upland rice yield and its corresponded projected values by SALTMED model in the first growing season (**a**) and in the second growing season (**b**) under SSP5-8.5 in 2080–2099

plant respiration, evapotranspiration, plant water, and nutrient losses, and decreased crop growth stages, leading to lower season length, lower water and nutrient use efficiency in rice (Sharma et al., 2022). In Egypt, Eid (2001) indicated that the Egyptian national rice production is expected to decrease by 11%, as a result of climate change effects in 2050.

Intercropping Legume Crops with Upland Rice

Intercropping between cereals and legumes has been an advantageous production practice. Intercropping with legume crops can contribute in the sustainable use of soil resources. The two species involved in the intercropping system have roots with different morphology, which facilitate nutrient uptake from different soil layers (Li et al., 2006). The legume crops are also able to fix atmospheric nitrogen and share part of this nutrient with the companion crop and in succession (Calegari & Carlos, 2014). Additionally, they increase soil organic matter content, promote nutrients cycling and alter soil biota, providing conditions for the development of arbuscular mycorrhiza, which promoting increases in grain yield (Amado et al., 2014; Calegari & Carlos, 2014).

Intercropping also can contribute in the sustainable use of water resources. In intercropping, the secondary and main crops are using the applied irrigation water to the main crop, which contributes to the efficient and complementary use of applied irrigation water (Qin et al., 2013; Yang et al., 2011). When the two intercrops have different rooting depth it contributes in increasing water absorption from different soil layers and avoiding losses through percolation and evaporation and increasing water infiltration rate (Rahman et al., 2017). Hu et al. (2015) found that intercropping systems used more water and enhanced water use efficiency by an average of 26%. Furthermore, Coll et al. (2012) explained that the improvement of water productivity of intercropped crops, compared to sole planting is due to higher water capture efficiency. Maitra et al. (2020) indicated that the great yields attained by the intercrops are only a consequence of low water losses.

Being irrigated as any regular field crop, upland rice can be a good candidate in an intercropping system as a main crop. Several benefits can be obtained from this system, namely maximizing land use and ensure higher crop productivity, profitability, and sustainability in upland rice production system and it is also gainful because of the reduced cost of weed control as the canopy of both crops maintain ground covers and prevent weed growth (Cagasan & Amarado, 2023).

Papong and Cagasan (2020) intercropped upland rice with either peanut or mung bean. They found that peanut and upland rice intercropping system attained higher yield of upland rice, than the obtained from upland rice and mung bean intercropping system. Furthermore, they found a slight increase in organic matter and exchangeable potassium in the soil. Whereas, Cagasan and Amarado (2023) indicated that intercropping peanut with upland rice, where peanut was planted two weeks ahead of upland rice resulted in the highest combined yield of both crops. Putra et al. (2018) intercropped soybean with upland rice under different planting densities of both crops and they found that intercropping 80:20 upland rice:soybean attained the highest upland rice yield close to 100:0 upland rice:soybean. Furthermore, Mugisa et al. (2020) reported higher values of LER reaching 1.5 obtained from intercropping upland rice with beans or upland rice with groundnut.

Conclusion

The future climate change not only will reduce the productivity of the cultivating crops, but also it will increase the required amount of water to irrigate these crops. Our analysis showed that the future climate change will negatively affect the water requirements and water productivity of paddy rice, depending on the studied future scenario. Our results also showed that under SSP1-2.6 in both time-intervals, the required water to irrigate paddy rice every 6 days to attain the same yield obtained under the field experiment will increase by 23%, and that will reduce water productivity by 18%. Whereas, under SSP5-8.5 in both time-intervals, 31% increase in the required irrigation water will be expected, which will reduce water productivity by 21%. Whereas, under application of deficit irrigation to paddy rice, namely irrigation every 12 days, the percentage of increase in the water requirements will be 20 and 25% under SSP1-2.6 and SSP5-8.5, respectively for both time-intervals. Moreover, water productivity will be reduced by 28 and 32% under SSP1-2.6 and SSP5-8.5, respectively for both time-intervals.

Nevertheless, the increase in the water requirements of upland rice under future climate change scenarios were lower than the values obtained for paddy rice using the same modeling approach and under the same studied future scenarios. Under SSP1-2.6, the increase in the applied irrigation water will be increase by 18, 20, 17 and 17% for farmer practice, and irrigation every 6, 9 and 12 days, respectively with losses in water productivity by 15, 15, 14, and 14%, respectively. Similarly, under

SSP5-8.5, the increase in water requirements will be 24, 25, 23, and 23% for farmer practice, and irrigation every 6, 9 and 12 days, respectively with losses in water productivity by 18, 19, 18, and 18%, respectively.

Our results also showed that the losses in the yield of upland rice will depend on the used scenario and the time interval. The reduction in the yield will be between 8–11% and 9–12% under SSP1-2.6 in 2060–2079 and 2080–2099, respectively. Whereas, under SSP5-8.5, the reduction in the yield will be between 8–13% and 8–12%, respectively. The above results suggest that the cultivation of upland rice will be more suitable than paddy rice cultivation to conserve water resources in the future.

References

Abd El-Naby, S. S. M., & El-Ghandor, A. M. A. (2022). Chemical control of grassy weeds in drill-seeded rice (Oryza sativa L.). *Menoufia Journal of Plant Production, 7*, 167–179.

Abd El-Wahed, M. H., & Snyder, R. (2015). Calculating sunshine hours and reference evapotranspiration in arid regions when solar radiation data are limited. *Irrig. Drain.* https://doi.org/10.1002/ird.1920.

Acharjee, T. K., Ludwig, F., van Halsema, G., Hellegers, P., & Supit, I. (2017). Future changes in water requirements of Boro rice in the face of climate change in North-West Bangladesh. *Agricultural Water Management, 194*, 172–183.

Ahmad, A., Ashfaq, M., Rasul, G., Wajid, S. A., Khaliq, T., Rasul, F., Saeed, U., Rahman, M. H., Hussain, J., Baig, I. A., Naqvi, S. A. A., Ahmad, S., Naseem, W., Hoogenboom, G., & Valdivia, R. O. (2015). Impact of climate change on the rice–wheat cropping system of Pakistan. In D. Hillel, & C. Rosenzweig (Eds.), *Handbook of climate change and agro-ecosystems: The Agricultural Modeling Intercomparison and Improvement Project (AgMIP) integrated crop and economic assessments* (pp. 219–258). Imperial College Press.

Akinbile, C. O. (2010). Crop water use response of upland rice to different water distribution under sprinkler irrigation system. *Advances in Applied Science Research, 1*, 133–144.

Alam Khan, M. (2012). Current status of genomic based approaches to enhance drought tolerance in rice (Oryza sativa l.), an over view. *Molecular Plant Breeding, 3*, 1–10. https://doi.org/10.5376/MPB.2012.03.0001

Allen, R. G., Pereira, L. S., Raes, D., & Smith, M. (1998). Crop evapotranspiration: Guideline for computing crop water requirements. FAO No. 56, pp 109–158.

Alou, I. N., Steyn, J. M., Annandale, J. G., & Van der Laan, M. (2018). Growth, phenological, and yield response of upland rice (Oryza sativa L. cv. Nerica 4®) to water stress during different growth stages. *Agricultural Water Management, 198*, 39–52.

Amado, T. J. C., Fiorin, J. E., Arns, U., Nicoloso, R. S., & Ferreira, A. O. (2014). Adubação verde na produção de grãos e no sistema plantio direto. In O. F. Lima Filho, E. J. Ambrosano, F. Rossi, & J. A. D. Carlos (Eds.), *Adubação verde e plantas de cobertura no Brasil: fundamentos e práticas* (vol. 2, pp. 81–125). Embrapa.

Anis, G., Hassan, H., Saneoka, H., & El Sabagh, A. (2019). Evaluation of new promising rice hybrid and its parental lines for floral, agronomic traits and genetic purity assessment. *Pakistan Journal of Agricultural Sciences, 56*, 567–576.

Arslan, A., Majid, G. A., Abdallah, K., Rameshwaran, P., Ragab, R., Singh, M., & Qadir, M. (2015). Evaluating the productivity potential of chickpea, lentil and faba bean under saline water irrigation systems. *Irrigation and Drainage*. https://doi.org/10.1002/ird.1912

Atlin, G. N., Lafitte, H. R., Tao, D., Laza, M., Amante, M., & Courtois, B. (2013). Developing rice cultivars for high-fertility upland systems in the Asian tropics. *Field Crops Research, 97*, 43–52. https://doi.org/10.1016/j.fcr.2005.08.014

Bauwens, A., Sohier, C., & Degré, A. (2013). Impacts of climate change on the hydrology and management of water resources in the Meuse basin (bibliographic summary). *Biotechnologie, Agronomie, Societe et Environnement, 17*(1), 76–86. (in French).

Bilgili, A. V. (2013). Spatial assessment of soil salinity in the Harran plain using multiple kriging techniques. *Environmental Monitoring and Assessment, 185*, 777–795.

Bouman, B., & Tuong, T. (2001). Field water management to save water and increase its productivity in irrigated lowland rice. *Agricultural Water Management, 49*, 11–30.

Bouman, B. A. M., Lampayan, R. M., & Tuong, T. P. (2007) Water management in irrigated rice: Coping with water scarcity. International Rice Research Institute.

Cagasan, U., & Amarado, J. (2023). Influence of intercropping on the growth and yield performance of upland rice (Oryza sativa L. var. zambales) and Peanut (Arachis hypogaea L.). *Research in Agricultural Sciences, 54*(1), 2–8.

Calegari, A., & Carlos, J. A. D. (2014). Recomendações de plantio e informações gerais sobre o uso de espécies para adubação verde no Brasil. In O. F. Lima Filho, E. J. Ambrosano, F. Rossi, & J. A. D. Carlos (Eds.), Adubação verde e plantas de cobertura no Brasil: fundamentos e práticas (vol. 2, pp. 451–478). Embrapa.

CAPMAS. (2020). Central Agency of Public Mobilization and Statistics. *Annual Statistical Year Book 2020.* Egypt.

Capra, A., Consoli, S., & Scicolone, B. (2008). Deficit irrigation: Theory and practice. In D. Alonso, & H. J. Iglesias (Eds.), *Agricultural irrigation research progress.* Nova Science Publishers, Inc.

Carroll, S. A., Le, K. N., Moreno-García, B., & Runkle, B. R. K. (2020). Simulating soybean-rice rotation and irrigation strategies in Arkansas, USA using APEX. *Sustainability, 2020*(12), 6822.

Ceccarelli, S., Grando, S., Maatougui, M., Michael, M., Slash, M., Haghparast, R., Rahmanian, M., Taheri, A., Al-Yassin, A., Benbelkacem, A., Labdi, M., Mimoun, H., & Nachit, M. (2010). Plant breeding and climate changes. *Journal of Agricultural Science, Cambridge, 148*, 627–637.

Chai, Q., Gan, Y., Zhao, C., Xu, H., Waskom, R. M., Niu, Y., & Siddique, K. H. M. (2016). Regulated deficit irrigation for crop production under drought stress. A review. *Agronomy for Sustainable Development, 36*(3), 2–21.

Cho, J., Ko, G., Kim, K., & Oh, C. (2016). Climate change impacts on agricultural drought with consideration of uncertainty in CMIP5 scenarios. *Irrigation and Drainage, 65*, 7–15.

Christy, T., Lisamarie, W., Mark, M., Fleck, J., & Linquist, B. (2018). Alternate wetting and drying decreases methylmercury in flooded rice, systems. *Soil Science Society of America Journal, 82*, 115–125.

Chu, G., Wang, Z., Zhang, H., Liu, L., Yang, J., & Zhang, J. (2015). Alternate wetting and moderate drying increases rice yield and reduces methane emission in paddy field with wheat straw residue incorporation. *Food and Energy Security, 4*, 238–254.

Coll, L., Cerrudo, A., Rizzalli, R., Monzon, J. P., & Andrade, F. H. (2012). Capture and use of water and radiation in summer intercrops in the south-east Pampas of Argentina. *Field Crops Research, 134*, 105–113.

Crusciol, C. A. C., Soratto, R. P., Nascente, A. S., & Arf, O. (2013). Root distribution, nutrient uptake, and yield of two upland rice cultivars under two water regimes. *Agronomy Journal, 105*, 237–247.

Ding, Y. M., Wang, W. G., Zhuang, Q. L., & Luo, Y. F. (2020). Adaptation of paddy rice in China to climate change: The effects of shifting sowing date on yield and irrigation water requirement. *Agricultural Water Management, 228*, 105890.

Dingkuhn, M., Laza, M. R. C., Kumar, U., Mendez, K. S., Collard, B., Jagadish, K., Singh, R. K., Padolina, T., Malabayabas, M., Torres, E., Rebolledo, M. C., Manneh, B., & Sow, A. (2015). Improving yield potential of tropical rice: Achieved levels and perspectives through improved ideotypes. *Field Crops Research, 182*, 43–59.

Djaman, K., Mel, V. C., Diop, L., Sow, A., El-Namaky, R., & Manneh, B. (2018). Effects of alternate wetting and drying irrigation regime and nitrogen fertilizer on yield and nitrogen use efficiency of irrigated rice in the Sahel. *Water, 10*(711), 20–25.

Dubey, A. N., Verma, S., Goswami, S. P., & Devedee, A. K. (2018). Effect of temperature on different growth stages and physiological process of rice crop—A review. *Bulletin of Environment, Pharmacology and Life Sciences, 7*(11), 129–136.

Eid, H. M. (2001). Climate change studies on Egyptian Agriculture. Soils, Water and Environment Research Institute. Agricultural Research Center, Egypt.

Elsadek, E. A., Zhang, K., Hamoud, Y. A., Mousa, A., Awad, A., Abdallah, M., Shaghaleh, H., Hamad, A. A. A., Jamil, M. T., & Elbeltagi, A. (2024). Impacts of climate change on rice yields in the Nile River Delta of Egypt: A large-scale projection analysis based on CMIP6. *Agricultural Water Management, 292*. https://doi.org/10.1016/j.agwat.2024.108673

Ewert, F. (2012). Opportunities in climate change? *Nature Clinical Practice Endocrinology & Metabolism, 2*, 153–154.

Eyring, V., Bony, S., Meehl, G. A., Senior, C. A., Stevens, B., Stouffer, R. J., & Taylor, K. E. (2016). Overview of the coupled model intercomparison project phase 6 (CMIP6) experimental design and organization. *Geoscientific Model Development, 9*(5), 1937–1958.

FAO. (2018). FAOSTAT database collections. Food and Agriculture Organization of the United Nations. Food outlook biannual report on global food markets. Rome. www.fao.org/faostat

Fereres, E., & Soriano, A. (2007). Deficit irrigation for reducing agricultural water use. *Journal of Experimental Botany, 58*, 147–159.

Fghire, R. W., Said, A., Fatima, I. A., Oudou, B. O., & Ragab, R. (2015). Response of quinoa to different water management strategies: Field experiments and SALTMED model application results. *Irrigation and Drainage, 64*(1), 29–40.

Fukai, S., & Mitchell, J. (2022). Factors determining water use efficiency in aerobic rice. *Crop and Environment, 1*(1), 24–40.

Fulu, T., & Zhang, Z. (2013). Climate change, high-temperature stress, rice productivity, and water use in Eastern China: A new superensemble-based probabilistic projection. *Journal of Applied Meteorology and Climatology, 52*, 531–551. https://doi.org/10.1175/JAMC-D-12-0100.1

Gaballah, M. M., Metwally, A. M., Skalicky, M., Hassan, M. M., Brestic, M., El Sabagh, A., & Fayed, A. M. (2020). Genetic diversity of selected rice genotypes under water stress conditions. *Plants, 10*, 27.

Golabi, M., Naseri, A. A., & Kashkuli, H. A. (2009). Evaluation of SALTMED model performance in irrigation and drainage of sugarcane farms in Khuzestan province of Iran. *Journal of Food, Agriculture and Environment, 7*, 874–880.

Hadi, A., Inubushi, K., & Yagi, K. (2010). Effect of water management on greenhouse gas emissions and microbial properties from paddy fields of Indonesia and Japan. *Paddy and Water Environment, 8*, 319–324.

Haefele, S. M., Jabbar, S. M. A., Siopongco, J. D. L. C., et al. (2008). Nitrogen use efficiency in selected rice (Oryza sativa L.) genotypes under different water regimes and nitrogen levels. *Field Crops Research, 107*(2), 137–146.

Hargreaves, G. H., & Samani, Z. A. (1982). Estimating potential evapotranspiration. *Journal of Irrigation and Drainage Division, 108*(3), 225–230.

Hargreaves, G. H., & Samani, Z. A. (1985). Reference crop evapotranspiration from temperature. *Transaction of ASAE, 1*(2), 96–99.

Hina, T., Adil, S. A., Ashfaq, M., & Ahmad, A. (2019). Economic impact assessment of climatic change sensitivity in rice-wheat cropping system of Pakistan. *Indian Journal of Science and Technology, 12*, 37.

Hirich, A., Ragab, R., Jacobsen, S. E., El Yousfi, L., & El Omari, H. (2012). Using deficit irrigation with treated wastewater in the production of quinoa (Chenopodium quinoa Willd.) in Morocco. *Revista Cientıfica UDO Agrıcola,* 12(3), 570–583.

Hong, E. M., Nam, W. H., Choi, J. Y., & Pachepsky, Y. A. (2016). Projected irrigation requirements for upland crops using soil moisture model under climate change in South Korea. *Agricultural Water Management, 165*, 163–180.

Hu, F., Chai, Q., Yu, A., Yin, W., Cui, H., & Gan, Y. (2015). Less carbon emissions of wheat–maize intercropping under reduced tillage in arid areas. *Agronomy for Sustainable Development, 35*, 701–711. https://doi.org/10.1007/s13593-014-0257-y

Husin, Y. A., Murdiyarso, D., Khalil, M. A. K., Rasmussen, R. A., Shearer, M. J., Sabiham, S., Sunar, A., & Adijuwana, H. (1995). Methane flux from Indonesian wetland rice: The effects of water management and rice variety. *Chemosphere, 31*, 3153–3180.

Jagadish, S. V. K., Craufurd, P. Q., & Wheeler, T. R. (2007). High temperature stress and spikelet fertility in rice (Oryza sativa L.). *Journal of Experimental Botany, 58*(7), 1627–1635.

Jamieson, P. D., Porter, J. R., Goudriaan, J., Ritchie, J. T., Keulen, H., & Stol, W. (1998). A comparison of the models AFRCWHEAT2, CERES-Wheat, Sirius, SUCROS2 and SWHEAT with measurements from wheat grown under drought. *Field Crops Research, 55*, 23–44.

Joshi, R., & Karan, R. (2014). Physiological, biochemical and molecular mechanisms of drought tolerance in plants. In R. K. Gaur, & P. Sharma (Eds.), *Molecular approaches in plant abiotic stress* (pp. 209–231). CRC Press.

Kadam, N. N., Tamilselvan, A., Lawas, L. M. F., Quinones, C., Bahuguna, R. N., Thomson, M. J., et al. (2017). Genetic control of plasticity in root morphology and anatomy of rice in response to water deficit. *Plant Physiology, 174*, 2302–2315. https://doi.org/10.1104/PP.17.00500

Kato, Y., Kamoshita, A., Yamagishi, J., & Abe, J. (2006a). Growth of three rice (Oryza sativa L.) cultivars under upland conditions with different levels of water supply: 1. Nitrogen content and dry matter production. *Plant Production Science, 9*, 422–434.

Kato, Y., Kamoshita, A., Yamagishi, J., & Abe, J. (2006b). Growth of three rice (Oryza sativa L.) cultivars under upland conditions with different levels of water supply: 2. Grain yield. *Plant Production Science, 9*, 435–445.

Kato, Y., Kamoshita, A., Yamagishi, J., & Abe, J. (2007). Growth of rice (Oryza sativa L.) cultivars under upland conditions with different levels of water supply 3. Root system development, soil moisture change and plant water status. *Plant Production Science, 10*, 3–13.

Kim, J., Shon, J., Lee, C. K., Yang, W., Yoon, Y., Yang, W. H., Kim, Y. G., & Lee, B. W. (2011). Relationship between grain filling duration and leaf senescence of temperate rice under high temperature. *Field Crops Research, 122*, 207–213.

Kögler, F., & Söffker, D. (2017). Water (stress) models and deficit irrigation: System-theoretical description and causality mapping. *Ecological Modelling, 361*(10 October), 135–156.

Kukal, S. S., Hira, G. S., & Sidhu, A. S. (2005). Soil matric potential-based irrigation scheduling to rice (Oryza sativa). *Irrigation Science, 23*, 153–159.

Kumar, A., Nayak, A. K., Pani, D. R., & Das, B. S. (2017). Physiological and morphological responses of four different rice cultivars to soil water potential based deficit irrigation management strategies. *Field Crops Research, 205*, 78–94.

Leng, G., & Hall, J. (2019). Crop yield sensitivity of global major agricultural countries to droughts and the projected changes in the future. *Science of the Total Environment, 654*, 811–821. https://doi.org/10.1016/j.scitotenv.2018.10.434

Li, L., Sum, J., Zhang, F., Guo, T., Bao, X., Smith, F. A., & Smith, S. E. (2006). Root distribution and interactions between intercropped species. *Oecologia, 147*, 280–290.

Lizcano Toledo, R., Lerda, C., Moretti, B., Miniotti, E., Santoro, V., Fernandez-Ondoño, E., Martin, M., Said-Pullicino, D., Romani, M., & Celi, L. (2022). Cover crops increase N and P cycling and rice productivity in temperate cropping systems. *Agronomy, 12*, 2193. https://doi.org/10.3390/agronomy12092193

Luo, Q. (2011). Temperature thresholds and crop production: A review. *Climate Change, 109*, 583–598. https://doi.org/10.1007/s10584-011-0028-6

Mahmoud, A., Ouda, S., & Abd El-Hafez, S. (2016). High water consuming crops under control: I. Case of rice crop. In *Major crops and water scarcity in Egypt* (pp. 69–82). Springer Publishing House. ISBN 978-3-319-21770-3.

Mahmood, S., Wahid, A., Javed, F., & Basra, S. M. A. (2010). Heat stress effects on forage quality characteristics of maize (Zea mays) cultivars. *International Journal of Agriculture and Biology, 12*, 701–706.

Maitra, S., Shankar, T., & Banerjee, P. (2020). Potential and advantages of maize-legume intercropping system. In A. Hossain (Ed.), *Maize production and use*. Intechopen. https://doi.org/10.5772/intechopen.91722

Marschner, H. (1978). Role of the rhizosphere in iron nutrition of plants. *Iran Journal of Agricultural Research, 6*, 69–80.

Miniotti, E. F., Romani, M., Said-Pullicino, D., Facchi, A., Bertora, C., Peyron, M., Sacco, D., Bischetti, G. B., Lerda, C., Tenni, D., et al. (2016). Agro-environmental sustainability of different water management practices in temperate rice agro-ecosystems. *Agriculture, Ecosystems & Environment, 222*, 235–248.

Mishra, A., Singh, R., Raghuwanshi, N. S., Chatterjee, C., & Froebrich, J. (2013). Spatial variability of climate change impacts on yield of rice and wheat in the Indian Ganga Basin. *Science of the Total Environment, 468–469*(Supplement), S132–S138. https://doi.org/10.1016/j.scitotenv.2013.05.080

Mohamed, A., Sedeek, S., Galal, A., & Alsakka, M. (2019). Effect of water deficiency as abiotic stress on the reproductive and ripening stage of rice genotypes. *International Journal of Agriculture and Plant Science, 2*(1), 13–19.

Mugisa, I., Fungo, B., Kabiri, S., Sseruwu, G., & Kabanyoro, R. (2020). Productivity optimization in Rice-based intercropping systems of Central Uganda. *International Journal of Environment, Agriculture and Biotechnology, 5*, 142–149. https://doi.org/10.22161/ijeab.51.22

N'guessan, K. J. Y., Adahi, B., Konan-Waidhet, A., Masayoshi, S., & Assidjo, N. E. (2023). Assessment of climate change impact on water requirement and rice productivity. *Rice Science, 30*(4), 276–293.

Nasir, I. R., Rasul, F., Ahmad, A., Asghar, H. N., & Hoogenboom, G. (2020). Climate change impacts and adaptations for fine, coarse, and hybrid rice using CERES-Rice. *Environmental Science and Pollution Research, 27*(9), 9454–9464.

Noreldin, T., & Abd El-Bakey, H. (2017). Response of some rice cultivars to different irrigation regimes in North Nile Delta, Egypt. *Fayoum Journal of Agricultural Research & Development, 31*(1), 162–173.

Oh-e, I., Saitoh, K., & Kuroda, T. (2007). Effects of high temperature on growth, yield and dry-matter production of rice grown in the paddy field. *Plant Production Science, 10*(4), 412–422.

Otengdarko, P., Kyeibaffour, N., & Ofori, E. (2012). Simulating rice yields under climate change scenarios using the CERES-rice model. *African Crop Science Journal, 20*, 401–408.

Ouda, S. (2020). Projected crop coefficients under climate change in Egypt. In *Climate change impacts on agriculture and food security in Egypt*. Springer Publishing House. ISBN: 978-3-030-41628-7

Pachauri, R. K., & Meyer, L. A. (2014). Climate change 2014: Synthesis report. In *Contribution of working groups I, II and III to the Fifth Assessment Report of the Intergovernmental Panel on Climate Change*, IPCC, Geneva, Switzerland, pp. 1–151.

Papong, J. R., & Cagasan, U. A. (2020). Growth and yield performance of upland rice (Oryza sativa L. var. zambales) intercropped with mung bean (Vigna radiata L.) and peanut (Arachis hypogaea L.). *International Journal of Life Science and Agricultural Research, 4*(1), 34–41.

Prasad, P. V. V., Boote, K. J., Allen, L. H., Sheehy, J. R. J. E., & Thomas, J. M. G. (2006). Species, ecotype and cultivar differences in spikelet fertility and harvest index of rice in response to high temperature stress. *Field Crops Research, 95*, 398–411.

Prell, J., & Poole, P. (2006). Metabolic changes of rhizobia in legume nodules. *Trends in Microbiology, 14*, 161–168.

Pulvento, C., Lavini, A., Riccardi, M., d'Andria, R., & Ragab, R. (2015). Assessing amaranth adaptability in a Mediterranean area of south Italy under different climatic scenarios. *Irrigation and Drainage, 64*(1), 50–58.

Pulvento, C., Riccardi, M., Lavini, A., D'andria, R., & Ragab, R. (2013). SALTMED model to simulate yield and dry matter for quinoa crop and soil moisture content under different irrigation strategies in south Italy. *Irrigation and Drainage, 62*(2), 229–238.

Putra, F., Yudono, P., & Waluyo, S. (2018). Growth and yield of upland rice under intercropping system with soybean in sandy coastal area. *Ilmu Pertanian (Agricultural Science), 2*(3), 130–136. https://doi.org/10.22146/ipas.25215

Qin, A. Z., Huang, G. B., Chai, Q., Yu, A. Z., & Huang, P. (2013). Grain yield and soil respiratory response to intercropping systems on arid land. *Field Crops Research, 144*, 1–10. https://doi.org/10.1016/j.fcr.2012.12.005

Ragab, R. (2002). A holistic generic integrated approach for irrigation, crop and field management: The SALTMED model. *Environmental and Modelling Software, 17*(4), 345–361.

Ragab, R. (2015). Integrated management tool for water, crop, soil and N-fertilizers: The saltmed model. *Irrigation and Drainage, 64*(1), 1–12.

Ragab, R., Malash, N., Gawad, G. A., Arslan, A., & Ghaibeh, A. (2005a). A holistic generic integrated approach for irrigation, crop and field management: 1. The SALTMED model and its calibration using field data from Egypt and Syria. *Agricultural Water Management, 78*(1), 67–88.

Ragab, R., Malash, N., Gawad, G. A., Arslan, A., & Ghaibeh, A. (2005b). A holistic generic integrated approach for irrigation, crop and field management: 2. The SALTMED model validation using field data of five growing seasons from Egypt and Syria. *Agricultural Water Management, 78*(1), 89–107.

Rahman, M. M., Islam, A. M., Azirun, S. M., & Boyce, A. N. (2014). Tropical legume crop rotation and nitrogen fertilizer effects on agronomic and nitrogen efficiency of rice. *The Scientific World Journal, 2014*, 11. Article ID 490841. https://doi.org/10.1155/2014/490841

Rahman, T., Liu, X., Hussain, S., Ahmed, S., Chen, G., & Yang, F. (2017). Water use efficiency and evapotranspiration in maize-soybean relay strip intercrop systems as affected by planting geometries. *PLoS ONE, 12*(6), e0178332. https://doi.org/10.1371/journal.pone.0178332

Rai, R. K., Singh, V. P., & Upadhyay, A. (2017). Irrigation methods. In *Planning and evaluation of irrigation projects* (pp. 353–363). Academic Press.

Razzaghi, F., Plauborg, F., Ahmadi, S. H., Jacobsen, S.-E., Andersen, M. N., & Ragab, R. (2011). Simulation of quinoa (Chenopodium quinoa Willd.) response to soil salinity using the SALTMED model. In *ICID 21st international congress on irrigation and drainage, 15–23 October 2011*, International Commission on Irrigation and Drainage, Tehran, Iran, pp. 25–32.

Rehman, M. U., Rather, G. H., Gull, Y., Mir, M. R., Mir, M. M., Waida, U. I., & Hakeem, K. R. (2015). Effect of climate change on horticultural crops. In K. R. Hakeem (Ed.), *Crop production and global environmental issues* (pp. 211–239). Springer International Publishing.

Runkle, B. R., Suvo, C. K., Reba, M. L., Reavis, C. W., Smith, S. F., Chiu, Y. L., & Fong, B. (2019). Methane emission reductions from the alternate wetting and drying of rice fields detected using the eddy covariance method. *Environmental Sciences Technology, 53*, 671–681.

Satake, T., & Yoshida, S. (1978). High temperature-induced sterility in indica rice at flowering. *Japanese Journal of Crop Science, 47*(1), 6–17.

Setyanto, P., Pramono, A., Adriany, T. A., Susilawati, H. L., Tokida, T., Padre, A. T., & Minamikawa, K. (2017). Alternate wetting and drying reduces methane emission from a rice paddy in Central Java, Indonesia without yield loss. *Soil Science Plant Nutrition, 64*, 23–30.

Sharma, R. K., Kumar, S., Vatta, K., et al. (2022). Impact of recent climate change on corn, rice, and wheat in southeastern USA. *Science and Reports, 12*, 16928. https://doi.org/10.1038/s41598-022-21454-3

Sheha, A. M., Abd El-Lattef, A. M., & Ouda, S. (2021) Decreasing maize production-consumption gap by intercropping with upland rice using different planting densities under deficit irrigation. *Moroccan Journal of Agricultural Sciences, 2*(4), 182–193.

Sheha, A. M., Ahmed, N. R., & Abou-Elela, A. M. (2014). Effect of crop sequence and nitrogen levels on rice productivity. *Annals of Agricultural Science, 52*(4), 451–460.

Singh, C. M., Kumar, B., Mehandi, S., & Chandra, K. (2012). Effect of drought stress in rice: A review on morphological and physiological characteristics. *Trends Bioscience, 5*, 261–265. https://doi.org/10.1016/j.heliyon.2023.e13744

Silva, L. L., Ragab, R., Durante, I., Lourenço, E., Simões, N., & Chaves, M. M. (2013). Calibration and validation of SALTMED model under dry and wet conditions using chickpea field data from southern Portugal. *Irrigation Science, 31*, 651–659.

Snyder, R. L., Orang, M., Bali, K., & Eching, S. (2004). Basic irrigation scheduling BIS. http://www.waterplan.water.ca.gov/landwateruse/wateruse/Ag/CUP/Californi/Climate_Data_010804.xls

Thuy, N., Shan, H., Bijay, S., et al. (2008). Nitrogen supply in rice-based cropping systems as affected by crop residue management. *Soil Science Society of America Journal, 72*(2), 514–523.

Valipour, M. (2014). Pressure on renewable water resources by irrigation to 2060. *Acta Advances in Agricultural Sciences, 2*, 32–42. http://www.aaasjournal.org

Vermeiren, L., & Jopling, G. A. (1984). Localized irrigation. FAO, Irrigation and Drainage Paper No. 36, Rome, Italy.

Vermeulen, S. J., Aggarwal, P. K., Ainslie, A., Angelone, C., Campbell, B. M., Challinor, A. J., Hansen, J. W., Ingram, J. S. I., Jarvis, A., Kristjanson, P., Lau, C., Nelson, G. C., Thornton, P. K., & Wollenberg, E. (2012). Options for support to agriculture and food security under climate change. *Environmental Science & Policy, 15*(1), 136–144.

Vibhuti, V., Shahi, C., Bargali, K., & Bargali, S. (2015). Seed germination and seedling growth parameters of rice (Oryza sativa) varieties as affected by salt and water stress. *Indian Journal of Agricultural Sciences, 85*, 102–108.

Wang, Z. Q., Zhang, W. Y., Beebout, S. S., Zhang, H., Liu, L. J., Yang, J. C., & Zhang, J. H. (2016). Grain yield, water and nitrogen use efficiencies of rice as influenced by irrigation regimes and their interaction with nitrogen rates. *Field Crops Research, 193*, 54–69.

Wassman, R. (2019). Environmental footprints of modernization trends in rice production systems 984 of Southeast Asia. *Environmental Science.* https://doi.org/10.1093/acrefore/9780199389419.013.230

Wei, X. R., Hao, M. D., Shao, M., & Gale, W. J. (2006). Changes in soil properties and the availability of soil micronutrients after 18 years of cropping and fertilization. *Soil and Tillage Research, 91*, 120–130.

Weinert, C., de Sousa, R. O., Bortowski, E. M., et al. (2023). Legume winter cover crop (Persian clover) reduces nitrogen requirement and increases grain yield in specialized irrigated hybrid rice system. *European Journal of Agronomy, 142.* https://doi.org/10.1016/j.eja.2022.126645

Welch, J. R., Vincent, J. R., Auffhammer, M., Moyae, P. F., Dobermann, A., & Dawe, D. (2010). Rice yields in tropical/subtropical Asia exhibit large but opposing sensitivities to minimum and maximum temperatures. *Proceedings of the National Academy of Sciences, 107*, 14562–14567.

Wihardjaka, A., Tandjung, S. D., Sunarminto, B. H., & Sugiharto, E. (2010). Emisi gas dinitrogen oksida pada padi gogorancah oleh pemberian jerami padi dan bahan penghambat nitrifikasi Penelitian Pertanian Tanaman Pangan 29, 144–151.

Willmott, C. J. (1981). On the validation of models. *Physical Geography, 2*, 184–194.

World Bank Group. (2021). Climate change knowledge portal (CCKP) user manual (18 p). World Bank Group.

Yadav, M. K., Aravindan, S., Ngangkham, U., Raghu, S., Prabhukarthikeyan, S. R., et al. (2019). Blast resistance in Indian rice landraces: Genetic dissection by gene specific markers. *PLoS ONE, 14*(3), e0213566.

Yang, C., Huang, G., Chai, Q., & Luo, Z. (2011). Water use and yield of wheat/maize intercropping under alternate irrigation in the oasis field of northwest China. *Field Crops Research, 124*, 426–432.

Yang, J. C., Liu, K., Wang, Z. Q., Du, Y., & Zhang, J. H. (2007). Water-saving and high-yielding irrigation for lowland rice by controlling limiting values of soil water potential. *Journal of Integrative Plant Biology, 49*, 1445–1454. https://doi.org/10.1111/j.1672-9072.2007.00555.x

Yin, Y., Tang, Q., & Liu, X. (2015). A multi-model analysis of change in potential yield of major crops in China under climate change. *Earth System Dynamics, 6*, 45–59.

Yoshida, S. (1981). *Fundamentals of rice crop science* (p. 269). IRRI, Los Baños.

Zeng, Y., & Li, F. (2020). Ridge irrigation reduced greenhouse gas emission in double-cropping rice field. *Archives of Agronomy and Soil Science*. https://doi.org/10.1080/03650340.2020.1771698

Zhang, Z. C., Li, H. W., Chen, T. T., Wang, X. M., Wang, Z. Q., & Yang, J. C. (2011). Effect of furrow irrigation and alternate wetting and drying irrigation on grain yield and quality of rice. *Science Agriculture Sin, 44*(24), 4988–4998.

Zheng, J. Z., Wang, W. G., Ding, Y. M., Liu, G. S., Xing, W. Q., Cao, X. C., & Chen, D. (2020). Assessment of climate change impact on the water footprint in rice production: Historical simulation and future projections at two representative rice cropping sites of China. *Science of the Total Environment, 709*, 136190.

Zou, J., Liu, A. L., Chen, X. B., Zhou, X. Y., Gao, G. F., Wang, W. F., & Zhang, X. W. (2009). Expression analysis of nine rice heat shock protein genes under abiotic stresses and ABA treatment. *Journal of Plant Physiology, 166*, 851–861.

SPRINGER NATURE

GPSR Compliance

The European Union's (EU) General Product Safety Regulation (GPSR) is a set of rules that requires consumer products to be safe and our obligations to ensure this.

If you have any concerns about our products, you can contact us on ProductSafety@springernature.com

In case Publisher is established outside the EU, the EU authorized representative is:

Springer Nature Customer Service Center GmbH
Europaplatz 3
69115 Heidelberg, Germany

The manufacturer's authorised representative in the EU is Springer Nature Customer Service Centre GmbH, Europaplatz 3, 69115 Heidelberg, Germany. If you have any concerns regarding our products, please contact ProductSafety@springernature.com

Printed and bound by CPI Group (UK) Ltd, Croydon, CR0 4YY
30/09/2026
02225590-0005